THE BIOLOGY OF DEATH

THE BIOLOGY of DEATH

ORIGINS OF MORTALITY

André Klarsfeld *&* Frédéric Revah

Translated from the French by Lydia Brady

COMSTOCK PUBLISHING ASSOCIATES

A DIVISION OF

CORNELL UNIVERSITY PRESS

ITHACA AND LONDON

The publisher gratefully acknowledges the assistance of the French Ministry of Culture— Centre national du livre.

First published 2004 by Cornell University Press

Printed in the United States of America

Library of Congress Cataloging-in-Publication Data

Klarsfeld, André.
  [Biologie de la mort. English]
  The biology of death/André Klarsfeld and Frédéric Revah ; translated by Lydia Brady.
    p. cm.
  Includes bibliographical references and index.
  ISBN 0-8014-4118-8 (cloth)
  1. Death (Biology) 2. Aging. 3. Cell death. I. Revah, Frédéric. II. Title.

QH530.K5613 2003
571.9'39—dc21

2003055169

... all we can hope for is that the boldness of the scientist should be tempered with scruples, that he never forget that—in the immortal words of Bacon—"science, if taken without the antidote of charity, cannot help but be somewhat malignant and poisonous to the spirit. "
—Jean Rostand, *Peut-on modifier l'homme?*

This warning could obviously have been addressed to the author of the lines below:

And even if we could prolong health until death is nigh, it would not be wise to give long life to all. We already know what problems arise when the number of individuals increases with no care being given as to their quality. Why increase the lifespan of people who are miserable, selfish, stupid, and useless?
—Alexis Carrel, *L'homme, cet inconnu*

In memory of all those whose lives were broken by criminal ideologies fostered by scientists with few scruples, like Carrel.

To our families.

CONTENTS

ACKNOWLEDGMENTS

Our book was born from the articulation, facilitated by the stimulating atmosphere of a dinner among friends, of a paradox: what could the life sciences have to say about death, which is the opposite of life? During the first phases of this project, we had frequent, long discussions with François Schächter, initiator of the Chronos project on the genetics of aging at the Centre d'étude du polymorphisme humain (CEPH) [Research Center of Human Polymorphism]. His vast knowledge in this area and his vision and constant commitment were extremely valuable to us. We present our warmest thanks to him.

Catherine Allais, Henri Atlan, Annick Barrau, Raymond Boudon, Jean-Charles Darmon, Pierre-Henri Gouyon, Dominique Guillo, Charles Lenay, and Michel Morange also helped us with advice, encouragement, and criticism at various stages of our reflection and during the actual writing process. We are very grateful to them. We also thank the Société de Thanatologie for its extensive bibliographical resources, and for allowing us to consult them so often. Two readers for Cornell University Press, Antonie W. C. Blackler and John Ewer, provided invaluable advice in preparing this English edition.

Above all, we would like to affectionately thank Florence and Isabelle, our wives, and our children, Marianne, Lucie, Benjamin, Raphael, and Elie, for their confidence and limitless patience. We hope that, in the end, they will find that it was all worth it!

PROLOGUE

How the Life Sciences Deal with Death

Death must be Copernicized.

E. Morin

Why devote an entire book to "natural" death? The reason for death seems clear in advance: death, the reader will say, the type that comes more or less directly from within the organism, is an absolute rule for living organisms, a consequence of aging, that slow degradation that is the inevitable mark of the passage of time. Any additional explanations seem superfluous, and the subject merits nothing more than a few consoling paragraphs. "Death is indeed useful," our common sense whispers to us, "it plays a regenerating role, it eliminates the oldest (read: 'the least useful') to benefit the young." Science writing for the general public often supports these conclusions. To give one example, Jacques Ruffié spoke of "the powerful selective advantage [of death], not so much at an individual level, but for the species" because "sexuality and accompanying death ensure change."[1]

However, even a superficial examination of the living world reveals the first fissures in this seemingly flawless vision. What about the incredible variety of life spans? The human species, whose most hardy representatives live beyond 120 years, is lucky compared with the fly, which lives at most a few weeks or months, not to mention certain mayflies, whose adult life can be counted in minutes. But we humans are nowhere near the record of millenarian sequoias or the less well-known but much more impressive wild blueberry bushes that have lived for 13,000 years! The repre-

[1] *Le sexe et la mort* (Paris, Odile Jacob, 1986).

sentatives of these species are all but immortal, and almost nothing but lightning or a lumberjack can kill them. Examples such as these suggest that death may not be programmed for all eternity into the very nature of the living, and that it is perhaps not an inescapable necessity. These examples also lead us to wonder why certain species seem to enjoy this exceptional good fortune, and how they managed to implement the biological mechanisms that allow them to resist time. For these species, a slogan like "make way for the young" decidedly does not apply!

The notion of absolutely necessary and universal death is not the only concept to show flaws on closer examination. As we examine the subject more closely, we will be led to challenge other preconceived notions, such as the existence of obligatory links between death and sexuality, and between death and the complexity of organisms. Simple organisms, which reproduce asexually, by fission or budding, can also age and die. Moreover, death is not the toll of the high degree of differentiation that typifies multicellular beings. Brewer's yeast, though it is single-celled, nonetheless follows an aging curve that is very similar to our own.

These examples show that a strictly biological approach to death is possible only if, while searching for justifications rather than explanations, we avoid any attempt to moralize. We will discover the fascinating contribution of evolutionary theory to this approach, specifically in the reflections of German biologist August Weismann, more than a century ago. The interest and richness of this theoretical framework did not begin to be recognized until the 1950s. Even today, despite its great coherence and the vast amount of experimental data supporting it, Weismann's theory remains unfamiliar to the general public, and even to scientists who work in other fields.

The study of the death of organisms cannot be conceived without an understanding of how cells, the building blocks of living beings, function. To illustrate this, let's take another look at the sequoia. Its trunk is composed of living cells only on the youngest, outermost layers. The inside, which is almost the entire tree, is composed of dead cells. Although a sequoia, as an individual, is several thousand years old, none of its cells survives more than a few decades. Is there is a link between the life cycle of cells and the life cycle of the organism? The debate on this issue has been particularly heated. Alexis Carrel, during the first half of the twentieth century, asserted that the origin of aging and death was not to be found within cells. He claimed to have demonstrated that all cells were potentially immortal, provided that they were removed from the organism they were originally part of.

This dogma of cell immortality was not struck down until the early 1960s, when Leonard Hayflick discovered cell senescence, which manifests itself by an intrinsic limit in the ability of cells to divide. At that point, it became legitimate to look for the mechanisms of aging in cells in culture. And in 1964, the expression "programmed cell death" was coined to describe a different phenomenon, resembling a concerted suicide of entire groups of cells. Many authors had already observed it, in passing, as early as the mid-nineteenth century. The persistent lack of interest in this topic doubtless had the same roots as the fascination exerted by the dogma of cell immortality: How could the cell, the building block of living organisms, have anything to do with death? It took another twenty years before researchers realized the importance of cell death—before they demonstrated that cell suicide benefited the organism. This is indeed a case of death working for the good of life, particularly in the course of embryonic development, or in the nervous and immune systems. But if cell death goes awry, it can also threaten life, through cancer, AIDS, and neurological diseases. During the last decade, an entire field of research has opened up, and the explosion has been such that, (half) in jest, it has been predicted that "Death Sciences Institutes" will be created. To date, however, it cannot be asserted with certainty that programmed cell death, any more than cell senescence for that matter, plays a direct role in the aging and death of individuals. But both clearly have their place in the global understanding of normal aging and age-related diseases.

The passage from the cell to the organism is sometimes invoked in support of one of the preconceived notions mentioned earlier. Just as cell death can be programmed, for the benefit of the individual, it is said that the death of individuals may also be programmed to serve a larger design. This type of thinking is very reassuring. But do the biological data support it? This is the issue we attempt to analyze.

Let's start down the path that leads to the "Copernicization" of death. After a brief overview of the beliefs of ancient times, our first stop will be Sweden in the eighteenth century, with the great naturalist Carl Linnaeus and his disciples.

CHAPTER ONE

THE BIOLOGY OF DEATH:
A Brief History

> The fact that death is often feared, like a monster we dare not look at
> square on, is undoubtedly one of the reasons for the ignorance of sci-
> ence in this area.
>
> Elie Metchnikoff

There is a "radical difference between physics and biology," according to the philosopher Georges Canguilhem: "Illness and death in the
living who have created physics, sometimes risking their lives, are not
problems of physics. Illness and death of living physicists and biologists
are problems of biology."[1] What subject could be more emotionally loaded
than death?

Long considered a divine malediction or punishment, death seemed
alien to living things. It was imposed on life, which was originally capable
of eternity. Authors of classical antiquity, such as Aristotle, said it clearly:
Lacking immortality, a privilege the gods have reserved for themselves,
humans can perpetuate through their descendents. Reproduction is a
means of escaping death to a certain extent, of communing with the gods.

Of the two boundary experiences of human life—namely reproduction
and death—it is on the latter that the founding myths are based. Death,
then, is excluded from a naturalist approach. This result is not surprising.
Death, especially violent death, is observed almost immediately. The sudden disappearance of a loved one induces people to reflect on the meaning
of life, much more than on the immediate or distant causes of death. In antiquity, there was little interest in studying the longevity of living species,

---

[1] G. Canguilhem, *Idéologie et Rationalité dans l'histoire des sciences de la vie* (Paris: Vrin,
1981), p. 138.

although that period produced very detailed, and sometimes fanciful, descriptions of methods of reproduction.[2] Nothing was too astonishing, nothing too marvelous, when it came to the engendering of the living, which indeed assumes extraordinarily diverse forms. After all, the relationship between the sexual act and the birth of a child is not obvious, and raises many questions. The mystery of death, on the other hand, is not so much death itself, which is inevitably commonplace, but its effect on the individual, for no one can report on the full experience.

The development of naturalism, in the eighteenth century, moved away from the view of ancient times by searching for a more rational justification for this imposed death which, of course, like all of creation, was still considered the fruit of divine will. Carl Linnaeus, the founder of biological classification systems, and his students provided the first and most complete attempt at rationalizing death. The son of a Swedish pastor, Linnaeus showed an early taste for botany, which led him to spend time studying plants at the expense of his schooling. Deemed unfit for serious studies, he became a cobbler's apprentice. Fortunately, he was later led into science by a local doctor who recognized his talent. According to Linnaeus, divine wisdom imposed a natural order, which was based on four related phenomena: propagation, geographic distribution, destruction, and preservation. Simple calculations showed that "even one plant, if left unchecked by animals, could cover and envelop our entire globe"[3] in a short time. This possibility, he reasoned, is why the "Sovereign Moderator" created predators, which "help . . . preserve a just proportion among all the species, thus preventing them from multiplying excessively to the detriment of man and animals."[4] Although "at first glance, we do not really admire the butchery and the horrible War of All against All,"[5] Linnaeus posited that all scourges—including contagious diseases, aging, even war itself—were created by God for the greater good of all living things, since there must be a balance between births and deaths. Prey was not created for the predator; on the contrary, the predator works in the service of the prey by preserving a just proportion in the balance of na-

[2] F. N. Egerton, *Isis* 59 (1968): 175–189; Pichot, A., *Histoire de la notion de vie* (Paris: Gallimard, 1993).
[3] C. Linnaeus, *Discours sur l'accroissement de la terre habitable*, 1744.
[4] C. Linnaeus, *Economie de la nature*, 1749.
[5] C. Linnaeus, *La police de la nature*, 1760.

ture. Without this, prey would be doomed to suffer famine—or even worse, it would eliminate other species by invading the planet.

According to the Linnaean school of thought, nothing was left to chance in the economy of nature, not even aesthetics: "So that fallen and dead trees do not remain useless to the Universe and no longer present such a sad sight, nature accelerates their destruction in a singular fashion. . . . How industriously nature works to destroy a single trunk!" Is it not admirable, for example, that: "the woodpecker, by pecking at rotten trees in search of insect larva, hastens their destruction so that they do not spoil the view for too long." The agents of destruction also had their role for the common good of all creation. If we were unable to fathom the utility of divine works, Linnaeus reasoned, it was only through ignorance of the designs that inspired them. God did not create anything in vain, even death.

The French naturalist Georges Louis Leclerc, Count of Buffon, contested the classification system proposed by Linnaeus, his exact contemporary. (They were both born in 1707.) Nor did he share Linnaeus's intransigent finalism. Buffon's ideas had a much more modern-day resonance when he admitted that nature is not exempt from mistakes—that it "tinkers," to use the term of François Jacob. But he proposed roughly the same ideas as Linnaeus on the fundamental balance of nature. Through the endless game of reproduction and death, Buffon reasoned, "the total quantity of life is always the same, and death, though it seems to destroy everything . . . does no harm to nature, which only shines the brighter. Nature does not allow death to annihilate species, but instead shows itself to be independent of death and of time by allowing individuals to be cut down and destroyed."

In any case, death did not occupy a preeminent position with respect to reproduction, as it did in ancient times. The two were considered equal—essential features of the living, and in permanent opposition and balance. Thus, death was no longer deemed an original, immutable center around which life organized itself as best it could. The "Copernicization" of death was underway.

Opposition and complicated balance were also present in the vitalistic conceptions that were in vogue at the beginning of the Age of Enlightenment. According to the vitalists, living beings owed their existence, and especially their survival, to the action of a "vital principle" that constantly struggled against physical properties. Physical properties were equated to forces of death. In living bodies, vital laws had to be present in order to oppose physical laws. Life was defined as a negation, a force that defied physical laws. No one expressed this principle better than the great anatomist Xavier Bichat, who founded the study of living tissues, or histology. At the very beginning of his *Physiological Research on Life and Death*, published in 1800,[6] Bichat coined the inescapable phrase that is still included in the definition of "Life" in most French dictionaries: "Life is the ensemble of functions that resist death."[7] The final outcome of the combat is played out in advance, because "it is the nature of vital properties to wear out."

It was the very year of Bichat's death, in 1802 (he was only thirty years old), that the word *biology* made its first appearance, penned by Jean-Baptiste de Monet, chevalier de Lamarck. He made a distinction between organic, "necessarily doomed to death," and inorganic, which was immortal since it was not alive. The temporal finiteness of organisms was even considered a primordial characteristic: "A living body is a natural body limited in its duration, organized in its parts . . . possessing what we call life, and necessarily doomed to lose it, that is, to suffer death, which is the end of its existence."[8]

We shouldn't think, however, that Lamarck and Bichat agreed about the status of death, though both considered it a programmed part of life. Lamarck rejected vitalism: "Nothing is more unlikely, and in fact, is less proven, than this supposed ability that one attributes to living bodies to resist the forces to which all other bodies are subjected."[9] He believed that living matter and raw matter were governed by the same physical laws. If these laws produced extremely particular results when applied to living matter, it was because of the extremely particular organization of living things. There is only one type of natural law, namely physical laws; only those circumstances under which they are implemented change. Lamarck's position was more philosophical than empirical[10] as he had no

---

[6] X. Bichat, *Physiological Research on Life and Death* (Paris: Flammarion, republished in 1994).

[7] This famous definition has a long history, since the words *defunct* and *function* share a common root. They come from the Latin *fungi*, which means "to perform, to accomplish." To paraphrase Bichat, the defunct has lost the function that allowed him to resist death, therefore he can no longer function. But he has also accomplished his life; he has no more functions to fulfill. He is deprived of the vital function, in the double sense of the word,— namely, both the ability and the reason to live.

[8] Manuscript unpublished until 1944: P.-P. Grassé, "Biologie," *Rev. Scientif.* 82 (1944): 267–276.

[9] J.-B. Lamarck, *Philosophie zoologique* (1809; repub., Paris: Flammarion, 1994).

[10] "Nature never complicates its methods unnecessarily." Lamarck, op. cit.

specific data on the particular way in which living things were organized, and even less on the origin of life, which remains one of the most difficult questions of biology. Lamarck did, however, reject the idea that the organism was constantly opposing forces of death that were purely external: "It is not true, as has been said, that everything surrounding living bodies tends to destroy them."[11] Lamarck refused to see only destructive conflicts between the organism and its environment. His reflections on the adaptation of species to their habitat led him, as we know, to formulate one of the first coherent theories of the evolution of living things. In the same text, he looked at the fundamental problem of the distinction between natural death and accidental death, stating that "the cause that essentially leads to the death of each living body is within it, and not outside it," because "it is the peculiarity of life to lead inevitably to death." Lamarck placed death directly within the living, rather than making it the inescapable final effect of murderous exterior forces that gradually replaced supposed vital forces. According to Lamarck, then, death was, if possible, even more inescapable.

Lamarck's path and posterity are somewhat poignant. In his *Esquisse d'une histoire de la biologie*,[12] Jean Rostand notes that "the personality of Lamarck, despite his great value as a botanist and zoologist, had something that inspired mistrust in rational men . . . In the area of chemistry, in particular, his opinions bordered on the ridiculous . . . he wanted to ignore the discoveries of Lavoisier, whom he combats with the naïve assurance of the self-taught." It is true that Lamarck, at the age of seventeen, had chosen to be a soldier! Quickly promoted to officer for his heroism in Germany, he was later seriously wounded and had to leave the army. In Paris, he dabbled in everything—banking, journalism, music, medicine, meteorology, botany, and the list goes on. His book *Flore Française* attracted the attention of Buffon, who had him appointed to the Jardin du Roi (the future Jardin de Plantes [Botanical Gardens]) in 1781. He became professor at the Museum of Natural History founded by the Convention, and in 1829 died blind, poor, and unknown. The successive reactionary regimes at the start of the century did not appreciate Lamarck, who had studied botany with Jean-Jacques Rousseau, and owed his professorship to the Revolution. From all points of view, "he came too late, he was an

---

[11] J.-B. Lamarck, *Recherches sur l'organisation des corps vivants* (1802; repub., Paris: Fayard, 1986). Here, Lamarck quotes Bichat.

[12] J. Rostand, *Esquisse d'une histoire de la biologie* (Paris: Gallimard, 1945).

eighteenth century biologist lost in the nineteenth century" according to the historian André Pichot.[13] With a little more distance, the influence of this scientist, who might have been considered the first true biologist, remains difficult to evaluate. His statue in the Jardin des Plantes in Paris was dedicated, in 1908 "to the founder of the doctrine of evolution," but his ideas on this subject are used today primarily as a foil to present Darwin's ideas, which have subsequently prevailed.[14]

There is no doubt, however, as to the influence of Claude Bernard, considered the founder of modern physiology. The "Bernardian Revolution" was a major scientific event of the second half of the nineteenth century, and Bernard died much decorated with honors. Bernard was a professor at the Collège de France and at the Museum of Natural History, as well as a member of the Académie Française and a senator. When he died in 1878, he was accorded a state funeral. There seems to be no comparison between this son of a winegrower, honored by the Second Empire and the republic, and Lamarck, the self-taught chevalier who disappeared virtually unnoticed, rejected by his contemporaries. It is even more striking to note that for both, the area in which they excelled was not the field that first attracted them. The young Lamarck wanted a military career, though his involvement in the military almost cost him his life. The young Bernard, at the time working as a pharmacist's assistant in Lyon in order to make ends meet, had a literary soul. He even wrote a tragedy, *Arthur de Bretagne*. After reading it, the critic Saint-Marc Girardin advised the novice playwright to consider another career. Bernard accordingly took up medical studies. It proved to have been good advice.

More seriously, although Bernard quoted Lamarck infrequently, like Lamarck, Bernard refused to support the idea that there was a hypothetical life force specific to living organisms, in constant opposition to death forces—forces that were purely physical and chemical and supposedly located in the external environment: "It is not at all by struggling against the cosmic environment that the organism develops and stays alive; it is, on the contrary, through adaptation, in accord with this environment."[15] Of course, adaptation is synonymous with effort. According to Bernard, the

[13] Pichot, op. cit.

[14] A. Langaney attempts to reestablish the balance in *La Philosophie biologique* (Paris: Belin, 1999).

[15] C. Bernard, *Leçons sur les phénomènes de la vie communs aux végétaux et aux animaux*, 1878.

organism strives constantly to preserve the strict internal conditions required for it to function properly. Thanks to sophisticated regulatory mechanisms, starting with those involved in nutrition, the organism manages to create and maintain its "internal environment" (to use Bernard's term). This environment, composed by and for cells, succeeds in being largely independent from external variations. By reacting to external variations, the organism cancels them out; by adapting, it becomes free.[16] It can be said, however, that because of this dynamic, organisms live both against and as a result of their environment. The environment, then, is simultaneously a promise and a threat: organisms "maintain their organization both because of and despite their openness to the exterior."[17]

To illustrate this point of view, Bernard used two aphorisms: "Life is creation," and "Life is death."[18] The second of these resolutely stated the exact opposite of Bichat's famous definition—that life is the ensemble of functions that resist death. Indeed, Bernard continued,

> in a living being, everything is created morphologically, organizes itself, and everything dies, destroys itself. . . . The organ is created. . . . On the other hand, the organs destroy themselves, disorganize themselves constantly, by their very processes. . . . The first of these two types of phenomena is unique, without direct analogy; it is particular, specific to living beings. The second, vital destruction, is on the contrary physical and chemical, most often the result of combustion, fermentation, putrefaction. . . . When we want to designate the phenomena of life, we are actually indicating the phenomena of death.

The manifestations of life studied by physiologists, Bernard concluded, consumed the living by the same processes at work "in the corpse after death."

This death that Bernard reintroduced into life was above all the symbol

---

[16] For example, think about so-called homeothermic animals—mammals and birds—whose body temperature does not change by more than a few tenths of a degree, whereas the external temperature range is one hundred times greater!

[17] Canguilhem, op. cit.

[18] We do not have room here to present the experiments on which Claude Bernard based his ideas. For a recent discussion of Bernard's concepts, see, for example, A. Prochiantz, *Claude Bernard, la révolution physiologique* (Paris: PUF, 1990).

of the physical and chemical laws that govern inert nature and leave behind a trail of damage and destruction. Similar processes are an integral part of life, even if they involve different procedures and conditions. "Vital destruction" and "organic creation" are both based on fermenting agents.[19] "Existence is . . . nothing other than a perpetual alternation of life and death, composition and decomposition. There is no life without death; there is no death without life." So what is the specificity of the living, apart from the intercession of particular molecules? It lies in the fact that organic creation is not only "chemical synthesis" but also "morphological synthesis." It obeys a rigorous plan to construct the organism from a fertilized egg during development of the embryo, and then to maintain the form despite a constant renewal of substance.

Bernard had no explanation for this plan, which he realized was transmitted from generation to generation. He even believed that it was impossible to study experimentally: "We can hardly contemplate vital morphology, since its essential factor, heredity, is not an element in our power, and thus we cannot control it as we do physical conditions and vital manifestations." Morphology, then, did not fall within the scope of physiology but rather within that of descriptive disciplines such as zoology and anatomy. What was most specific to life, he claimed, eluded scientific method.

Perhaps it is not surprising that Bernard, having decided not to deal with the development of an organism, did not say much about its death. Like his predecessors, he considered obsolescence and death to be general characteristics of living beings, much as he considered embryonic development to be a characteristic of living organisms.[20] But to say that death was intrinsic to the living did not explain it any more than the reference to the "vital principle" by Bichat and the vitalists explained life. Their principle was not an explanation, but simply an observation. Death was apparently just the negation of life, to which it seemed to cling by the force of things, like the dark face of a shiny medallion. When life stopped, how could biology—"the science that studies life and living beings while they are living"[21]—not be reduced to silence?

---

[19] Or, in modern terms, biomolecules that are enzymes.

[20] Obsolescence, illness, and death were grouped together. The three remaining characters were organization, generation, and nutrition (Bernard, C. *Leçons sur les phénomènes*, op. cit.)

[21] Pichot, op. cit.

One of the first people to describe the conditions for an experimental approach to natural death and its causes—and to tackle the issue seriously—was the French biologist (also of Russian origin) Elie Metchnikoff.[22] In 1903, Metchnikoff coined the term *gerontology* to describe the scientific study of aging, as well as the term *thanatology* (which did not take hold as successfully), to describe the scientific study of death.[23] According to Metchnikoff, this area of study was becoming urgent because "with the progress of medicine . . . future illnesses will not have the same magnitude that we currently see. . . . The problem of truly natural death, as the end of normal life, will become extremely important." A disciple of Pasteur, Metchnikoff worked primarily on infectious diseases. The complete eradication of smallpox, declared in 1980 by the World Health Organization, proved that his efforts were partly justified. The definition of "truly natural death," however, is still not clear, and the distinction between aging and illness is becoming increasingly important.[24]

Metchnikoff posed several essential questions: What exactly do we mean by "natural death," especially for species capable of living a long time? Is there is a universal mechanism of natural death? What experimental system will enable us to study death without risk of confusion with other processes, in particular diseases? Unconvinced by observations made on elderly humans or higher order animals, in whom the changes could just as well result from disease as from an internal process of natural death, Metchnikoff preferred to study "beings whose organization is incompatible with prolonged life," in particular insects such as the silkworm moth. In these moths, death is inevitably natural as long as they are protected from trauma and infection.

But it is incontestably to August Weismann that we owe the most serious investigations, starting in 1881, into the questions that are at the heart of our discussion: If the limitation of existence is not programmed from the outset in the structures of the living, where does it come from? Why, in most species, are living organisms, or even their cells, not immortal? What

[22] He won the Nobel Prize in Medicine in 1908 for his work in immunology, mostly carried out at the Institut Pasteur after anti-Semitic persecutions compelled him to leave Odessa in 1882.

[23] From the Greek roots *geron* ("old man") and Thanatos ("God of Death"). E. Metchnikoff, *Etudes sur la nature humaine*, 1st ed. (Paris: Maloine, 1903).

[24] See chapter 2.

is the biological significance of death? The evolution of Weismann's responses to these questions provides a helpful guide on a path strewn with confusion and paradox. This German biologist, basically unknown to the general public, is considered by some geneticists as one of the greatest biologists of all time, for his conceptual contributions in the fields of heredity and the evolution of species.

Born in Frankfurt in 1834, and showing an early enthusiasm for natural science, Weismann first went into medicine, which his parents felt would enable him to earn a good living. He hardly practiced as a doctor but instead quickly devoted himself to biology, becoming a professor at the University of Freiburg in 1866. Starting at about that time, a visual impairment gradually prevented him from using a microscope. Despite his great talent as an experimenter, however, this disease was perhaps providential, since it compelled him to devote himself to theoretical research on the essential problems of biology.

Weismann refused to consider death as a simple established fact, in the very nature of life:[25] "We cannot see why the cells do not have an infinite capacity to multiply, which would allow the organism to live forever. Likewise, from a purely physiological standpoint, we do not see any reason why the organism should not be able to function forever."[26] Unlike reproduction, said Weismann, "death is not an essential attribute of the living substance." Weismann thus completed the reversal of the ancient perspective. He claimed that the true primary phenomenon, without which life was not even conceivable, was reproduction, whereas death was not an essential part of life. Indeed, there are numerous examples that support his claim. Countless large trees have nothing to fear but lightning and chainsaws. If well maintained, bacterial cultures, as well as colonies of certain simple animals that divide by fission, seem eternal.

Accidental or artificial death due to unfavorable external conditions may, however, be considered inevitable in the long term. This outcome, however, is not what interested Weismann. Rather, it was "normal" or

[25] We quote Weismann several times in this chapter, in particular from translations of his two principal articles on death, "The Duration of Life" (1881) and "Life and Death" (1883). The English translation was published in *Essais sur l'hérédité et la sélection naturelle [Essays upon heredity and kindred biological problems]* (Oxford, UK: Oxford University Press, 1889).

[26] Weismann, op. cit.

"natural" death due to "purely internal causes, programmed in the organization itself, as the normal end of life."[27] While acknowledging that
"one of the most thorny problems of all physiology [is] knowing what is
the cause of death," he refined the distinction between external and internal causes, stressing that with age, "certain changes in the tissues damage
their ability to function . . . and end up leading directly to what we call
normal death, or result indirectly in death, by rendering the organism incapable of resisting harmful external influences of little importance."[28]
Even in the second case, the final outcome is not totally accidental, because the organism has in fact paved the way for its own death, although
without specifying the date or the means.

Giving a Meaning to Death (Why Does Death Exist?)

Weismann's work fits directly into the framework of Charles Darwin's
theory of evolution by natural selection, proposed in 1858.[29] It was in this
framework that Weismann wanted to explain the hereditary characteristics specific to each living species. Longevity is clearly one such characteristic. Why are the life spans of an elephant and a mouse so different? Initially, Weismann sought to understand natural death in terms of direct
biological utility, of primary adaptation: "It is only from the point of view
of *utility*[30] that we can understand the necessity of death." And a bit further: "I consider that death is not a primary necessity, but that it has been
secondarily acquired as an adaptation. I believe that life is endowed with
a fixed duration, not because it is contrary to its nature to be unlimited,
but because the unlimited existence of individuals would be a *luxury*
without any corresponding advantage."[31] Since obviously every individual would prefer not to die, the disappearance of the oldest individuals
could be useful only to the population or the species to which they belong.
Weismann's reasoning, in 1881, was appealing: even animals that are

---

[27] Weismann, op. cit.

[28] Weismann, op. cit.

[29] To the point, according to J. Rostand (op. cit.), "of being more Darwinist than Darwin,
and radically expelling any Lamarckian residue from the doctrine of the master." We will
return to this in chapter 4.

[30] All italics here added by us for emphasis.

[31] Weismann, op. cit.

potentially immortal nevertheless suffer wounds and injuries that progressively reduce their capacities; they are less and less capable of procreating. It is therefore "necessary that they be continually replaced by new
more perfect individuals. . . . From this follows, on the one hand, the necessity of reproduction, and on the other, the utility of death. Worn out individuals are not only valueless for the species, but they are even harmful,
for they take the place of those which are sound."[32] We cannot conceive of
a life that is perpetuated without reproduction, since accidents of all kinds
eliminate living beings little by little.[33] On the other hand, "natural" death
is not indispensable. However, because it is programmed into most
species, we must try to figure out why it exists—or more accurately, why
it appeared.

This first path explored by Weismann was flawed because of its circu-
larity, which many authors (including Metchnikoff) criticized. How can
we justify the appearance of a process that eliminates aged individuals?
By the advantage provided by eliminating those who are the least capable. But admitting that the "old" are inevitably less capable than the
"young" is simply to point out the effect being explained—that age is accompanied by a decrepitude that leads to death—without explaining the
cause. The initial hypothesis postulates that the "young" are in better condition anyway: They will take the place of their elders without the need to
call on a mechanism dedicated to this end, because the "old" are already
disadvantaged when it comes to dangers of all sorts and the competition
waged by their younger fellow creatures.

What Good Is Immortality? (Why Should Death Not Exist?)

Weismann in fact had done nothing more than shift the question. There
was still no explanation as to why individuals shouldn't be able to completely repair most of their injuries, which is just a way of reformulating
the original question, why do the capacities of individuals generally de-

[32] Weismann, op. cit.
[33] Perhaps life should even be considered as a particular mode of molecular organization
whose primary consequence is the preservation of information through time, despite the
disappearance of its successive material incarnations.

crease with age? After all, many organisms have considerable powers of renewal and regeneration: look at the legs of certain frogs, or simply our skin's ability to heal. Why aren't these capacities more extensive, or more effective? And why do they often decline with age, notably faster in certain species than in others?

Weismann realized that these first attempts to provide an answer were insufficient. His ideas on aging and death evolved considerably over the thirty years that followed his first attempts, in 1881.[34] He was definitely on the right track when he suggested that, if an unlimited ability to reproduce was not indispensable, it was extremely likely to be lost. When could this occur? Never in the simplest organisms, such as bacteria, which multiply by dividing into two equal parts. Either both "offspring" have an unlimited ability to reproduce, or neither has. In the latter case, however, the species would not survive long. It is clear that a bacterium can at no time, through the generations of its descendants, survive without the ability to reproduce. The loss of this ability could be due only to an ultimately fatal accident.

This scenario is completely different for multicellular organisms. As Weismann wrote in 1883, most of a multicellular organism's body may be viewed as nothing but "a secondary appendage of the real bearer of life—the reproductive cells."[35] Only these cells "are potentially immortal, in so far as they are able, under favorable circumstances, to develop into a new individual, or in other words, to surround themselves with a new body (soma)."[36]

In more erudite terms, Weismann distinguished between two major classes of cells: "germ" cells, which are directly involved in the reproduction of the organism, and "somatic" cells (from the Greek *soma*, which means "body"). This distinction between germ and somatic cells is the

[34] Unfortunately for him, and doubtless for the evolution of this entire area of biology, Weismann never explicitly abandoned his initial hypotheses, although he did admit to "certain errors of interpretation" on which he did not expound. This provides a glimpse into the evolution of his thinking, into the "stages in his research," because of the fact that he did not rectify later editions of his works; see T. Kirkwood and T. Cremer. *Human Genetics* 60 (1982): 101–121. Weismann thus helped perpetuate the misunderstanding of his ideas and prevented his contemporaries and successors from perceiving the full depth and originality of his later ideas. We attempt to do him justice in the present work.

[35] Weismann, op. cit.

[36] Weismann, op. cit. The "favorable circumstances" are fertilization and everything that follows.

foundation for Weismann's best-known contribution to biology, involving the science of heredity, which was still in its infancy at that time. In this work, we look only at the prospects that this distinction opens for understanding aging and natural death in living organisms.

The key to the problem was recognizing that all that matters, to ensure the succession of generations, is the immortality of germ cells. The rest of the body, the soma, need not be eternal. It is not eternally indispensable: it could in theory disappear as soon as it has transmitted its eggs, or its sperm or pollen, thereby producing at least one descendant and bringing it to reproductive maturity to ensure the continuity of the line. The immortality of the soma is accessory. But does this explanation show how immortality was lost?

In chapter 4 we examine the arguments that allowed Weismann, and many authors after him, to respond to this question in the affirmative. For now, let's look more closely at what is meant by the potential "immortality" of the reproductive cells. The word *immortality* has a specific meaning in modern cellular biology: It designates not eternal survival but rather the capacity for unlimited cell division. When we say that germ cells are immortal, we recognize the continuity of the cell line—which, from the fertilized egg, forms the sex cells that in turn provide the potential starting point for the next generation. This immortality, however, is only potential and partial. It is potential because no individual of any species can be absolutely sure of reaching reproductive maturity, and then of finding a sexual partner (which in general is necessary) and finally leaving a viable descendant that itself reaches reproductive maturity. It is partial, because in each generation (except in certain modes of asexual reproduction), there is a combination of the maternal line and the paternal line. Whatever the case may be, this specific immortality of the germ line is indispensable not only for the preservation of the species but also for the preservation of life. As Weismann put it, somewhat poetically, life "since . . . its first appearance upon the earth, in the lowest organisms, has continued without break; the forms in which it is manifested have alone undergone change. Every individual alive today—even the very highest—is to be derived in an unbroken line from the first and lowest forms."[37]

[37] Weismann, op. cit.

Death of Cells versus Death of the Individual

By insisting on the role of cells in the aging and death of organisms, Weismann laid the groundwork for research that continues today. However, the link between cell death and the death of the organism has proven to be more complex than he imagined, and remains a controversial issue.[38] A simple diagram can illustrate the complexity of the problem. It is based on two pairs: life/death and cell/organism. Between the death of the organism and the death of its cells, the relationship seems quite obvious: the former results in the latter, immediately or shortly thereafter. Most cells are often in good condition at the time of an organism's death, but when the organism dies, so do they. In Weismann's period, cells were recognized as the basic unit of all living beings.[39] Bernard had already suggested that "that which dies, like that which lives, is definitively the cell." However, he believed that cell death in general was the consequence only of an injured, sick, or aging organism's inability to maintain the composition of its internal environment. This rule had certain nuances. The death of the individual could sometimes be attributed to the death of certain essential cells, such as cardiac or nerve cells. But Bernard supposed that death was above all a global phenomenon that was not based on the autonomous functioning of the cells. Usually, the cells simply suffered the consequences of death.

Weismann, on the other hand, emphasized the inverse relationship between cell death and the death of the organism. He started from the observation that "the cells that form the living tissue base wear out as a result of their activity and function." Worn-out cells are normally replaced to maintain the integrity of tissues, which are thus constantly being renewed; but the cells themselves ultimately wear out because "the ability of the cells of the body to multiply by dividing is not infinite, but limited."[40] This fact does not mean that the immediate cause of death is the halt of cell multiplication. Simply a slowdown can disrupt the replacement of worn-out cells. Certain vital functions are no longer being ensured, long before the final limit of cell division is reached. In any case, Weismann's hypothesis led him to see "the natural death of an organism

---

[38] See chapter 5.
[39] The cell theory of Schleiden and Schwann was proposed in 1839.
[40] Weismann, op. cit.

[as] the termination—the hereditary limitation—of the process of cell division, which began with the segmentation of the ovum."[41] Death is the final step in a life plan that starts with fertilization; it is written into the fate of each cell.

Weismann considered that death was programmed in multicellular beings through an intrinsic limitation in the ability of their somatic cells to multiply. That limitation was made possible by "the division of work" introduced by the first differentiation processes between somatic cells and germ cells. Only germ cells must imperatively be immortal. Furthermore, he suggested that the maximum size of the individuals of a given species could be determined in the same way as their maximum longevity, reducing "the limitations of the organism in both space and time to one and the same principle."[42] Is it not necessary, he reasoned, to regulate the natural ability of cells to proliferate for the organism to maintain a size compatible with the physiological and physical constraints affecting it? Not the least of which is gravity: an elephant-sized bird would certainly not be able to fly!

In 1932, Bidder reiterated Weismann's suggestion on size control in adult organisms, observing a correlation between the indefinite growth of certain animals and their apparent absence of aging.[43] The animals in question are often aquatic species, such as fish (e.g., scorpion fish, plaice, sturgeon), lobsters, certain mollusks, and perhaps some amphibians, all of which are less subject to the constraints of gravity. Any swimmer can appreciate the sensation of weightlessness in water, described by Archimedes' principle. Likewise, if the size and longevity of numerous plant species do not seem intrinsically limited, it may be because these plants do not have to fight against the earth's attraction in order to move about. Bidder proposed that aging and natural death in most species are simply side effects of the regulatory mechanisms that have been designed to limit growth—to prevent growth in excess of the maximum tolerable size at adulthood. Once the adult size is reached, these same mechanisms end up limiting the cell division required simply to renew the tissues. A

[41] Weismann, op. cit.

[42] Weismann, op. cit.

[43] G. P. Bidder, *British Medical Journal* (1932): 583–585. Note that the absence of aging and indefinite growth are impossible to establish with certainty, since this would require an infinite period of observation. It is better science to talk about negligible or undetectable aging, along with growth that seems potentially unlimited.

mechanical image can be used to illustrate this idea. The tendency to grow is an accelerator, and brakes are supplied to compensate. Once growth has stopped, the brakes "regrettably" remain on, and their action ultimately prevails, definitively halting the machine and leading to death.

This tension, this fragile balance between the brakes and the accelerator, is indeed found within the cells, in the complex interactions that favor cell proliferation (oncogenes) and those that block it (anti-oncogenes). At the organism level, however, there is hardly any experimental support for Bidder's conjecture today. It is not the size of organisms that directly determines their life span. The proposed link between indefinite growth and longevity has nevertheless provided rich teachings, and contemporary biology has confirmed most of Weismann's intuitions on cell aging and its role in aging in general,[44] although some controversy remains on this issue.

We cannot present the history of these ideas without mentioning the French surgeon Alexis Carrel, Nobel laureate for medicine in 1912, who that same year published an article titled "On the Permanent Life of Tissues outside Organisms." The repercussions from publication of this work were immense. In it, Carrel directly contradicted the theories of Weismann[45] after discovering that cells removed from a chicken survived and reproduced in culture much longer than the life span of the animal itself. It was not until 1961 that Carrel's work was questioned seriously, and longer still before the likely origin of his error was recognized. Rewriting history is always dangerous. And yet, if Weismann's ideas had been accepted as they deserved to have been, the biology of aging might have progressed much more rapidly. Carrel's "demonstration" seriously discredited Weismann's theories, thus diverting biologists from the study of cell senescence. The theoretical, even conjectural nature of Weismann's work also gave rise to skepticism, at a time when experimental biology was seeking, not without difficulty, a concrete—and manipulable—substrate of heredity. History, with a capital H, could also have played a role. When World War I broke out, only two years after Carrel's initial publica-

---

[44] See chapter 5. In an attempt to avoid confusion, we use the word *aging* only for organisms, and the word *senescence* only for cells.

[45] Jean Rostand (op. cit.) comments in a footnote that Carrel's results are "contrary to Weismann's idea of an essential distinction between somatic cells and germ cells."

tion, Weismann, as an ardent German patriot, gave up all the scientific distinctions awarded by the "enemy." He died that same year. His attitude and the fact that his articles were originally published in German may also have alienated non-Germanists.

Up to that point, the fact that cells wore out and disappeared was mentioned only to explain why individuals that were formed by cells deteriorated and died—or, inversely, cell death was considered a consequence of the individual's death, which progressively led to the metabolic shutdown of all its cells. But let us return to the two pairs proposed earlier. There is theoretically a conceptual quasi-equivalence between cell death and organismal death, as the death of an organism's cells is at once inevitable and inevitably negative for the individual. Cell death is apparently related only to pathological phenomena, or is only a marginal phenomenon in normal individuals. All that remains is to acknowledge the same equivalence between cell life and organismal life, and our diagram is complete.

It turns out that such a simple diagram is radically wrong. In fact, cell death plays a crucial role in the life of the individual, both during embryonic development and in adults. Symmetrically, the uncontrolled survival of a cell that should have died may place the organism in danger of dying.[46] The best known example of this is cancer, but there are others. The life and death of cells, and of the individuals they form, in fact have a much more subtle relationship than we had imagined. It has taken a long time for this subtlety to impress itself on the scientific community. Despite the many prior experimental indications pointing to the physiological role of cell death,[47] researchers had trouble recognizing the scope of this role; they showed even more resistance to this idea than to the concept of cell senescence. It is interesting to compare this change in thinking to the attitude concerning the death of individuals. Many people—biologists and others—have come up with ingenious explanations (though without any tangible proof, as we will see in later chapters) asserting that the death of individuals provides some potential advantage to the species. And al-

[46] The physiological and pathological aspects of cell death are covered in chapters 6 and 7, respectively.

[47] Weismann's own very first works were on the massive cell death that accompanies the formation of adult insects in the cocoon. But he seems not to have connected this phenomenon with his own ideas on the evolution of natural death.

though indications regarding the utility of cell death were there all the while, researchers long refused to consider this possibility, since it did not fit in with their intuitive diagram of the cell as a basic unit of living things. They simply could not believe that a cell could be deliberately programmed to die.

CHAPTER TWO

THE DIFFICULT MEASURE OF BIOLOGICAL TIME

Or, Getting Old Is Like Dying a Little . . . More

The reader may be surprised that we have to devote an entire chapter to defining aging. Didn't we say that our main topic was death? And after all, doesn't everyone have ample indirect or direct personal experience with aging every day? But let's take a look at the sometimes unexpected connections between death and aging.

Defining aging turns out to be a much more difficult exercise than one would first imagine. Getting old is easy, one might say; you just have to live long enough. Accelerated-aging diseases, such as progeria or Werner's syndrome, reduce the wait to just a few years. They show a sad and blindingly evident caricature of the ravages of time. The effects of aging are also evident in ninety- or one-hundred-year-olds, who are the subjects of many genetic studies on longevity.[1] Even when people that age are in very good health, they are indubitably old. Their exceptional longevity seems, however, to be proof that they have aged less, or better, or more slowly, than average. But this "measurement" is taken after the fact, when almost the entire age category has passed away. To say that the survivors are more resistant to aging is tautological. It must be possible to define the aging of an individual other than by comparison with those who are already dead! We will see that this task, which appears so simple, has up to now proven to be impossible.

---

[1] We return to these two extremes of the human aging spectrum in chapter 5.

Nothing measures the aging of an organism more accurately than the passage of time.

Aging and the Passage of Time

In some languages, such as English and German, people start getting old literally at birth, as in the question "How old are you?" The passage of time certainly has something to do with aging and death. To clarify this connection, let's start with the simplest living beings, bacteria. They reproduce by symmetrical division to create new identical cells. In this case, how can we talk about aging, since there is no individual[2] whose existence can be followed over time beyond the few hours, sometimes even minutes, that a generation lasts? Only distinct, unequivocally identifiable individuals can age and die. As François Jacob stated, "bacteria do not die. They disappear as an entity: where there was one, there are suddenly two."[3] It is difficult to conceive of death that leaves behind no corpse.

Still, the notions of death and age may have a meaning even for bacteria. Antibiotics and antiseptics are capable of killing bacteria. If a bacterium finds no more food, it ends up lysing—literally, it empties its contents, and its membrane dissolves. This is the clearest, most irremediable sign of the death of a cell. Also, certain bacteria have the ability to form a spore by a particular type of division. The spore is a resistant version of the bacterium that can survive a lack of food or extreme temperatures. It resumes multiplying as soon as the situation improves. How long can this lethargy last? The discovery in 1995 of bacterial spores twenty-five million years "old" suggests that there is virtually no limit other than that imposed by the physical preservation of the spore. Even if this spectacular result has not been confirmed, it gives us the opportunity to ask ourselves about the use of the word *old*. Their life having been suspended, the spores were comparable, for all these years, to a few fragments of inert matter. They are old in the same way that mountain ranges (like the Appalachians) or fossils are old: They have not really aged in the biological sense of the word. It would be better to say that they are ancient, and that they have been brought back to life, because they did not really live all

[2] Etymologically, meaning "that which does not divide."
[3] F. Jacob, *La Logique du vivant* (Paris: Gallimard, 1970), p. 317.

those millions of years.[4] Their resurrection is just as great a feat as their ancientness. Death is defined here by the inability to bring the spores back to life. The more time that has elapsed, the lesser the likelihood that they would survive more or less intact, still able to produce living bacteria.

Everyone knows that the passage of time is accompanied by spontaneous changes that generally move in the direction of destruction and disorder. Physicists have provided a rigorous foundation for this intuition by developing a precise measurement for disorder, called entropy. The second law of thermodynamics says that a closed system can move only to a higher state of entropy (or, at best, remain stable, but only under certain unrealistic conditions that need not be discussed here). In much less scientific and less abstract terms, any set of objects, whether they be atoms, marbles, or galaxies, seek to occupy all the space available to them, taking into account the interactions that exist among the objects. Any ordered structure within this set ultimately disappears. This principle is how classical physics defines the arrow of time. The more time passes, the lesser the likelihood that a bacterial spore (or a fossil, or a treasure) will be found intact. This result has nothing to do intrinsically with the spore or the fossil or the treasure, except in terms of its initial ability to passively resist the ravages of time.

Aging and Mortality

The complex arrangement of a living cell, not to mention a multicellular organism, constitutes an obvious challenge to this tendency toward disorder, which is largely identical to the passage of time. Of course, this very arrangement of living systems is what allows them to escape the harsh law of entropy because it gives them the ability to exchange matter and information with their environment. Living organisms are open systems; they are capable of incorporating external elements, renewing their substance—which, if we look closely, is in fact much "younger" than the organism itself. Still, with age, the exchanges become less and less efficient; the renewal slows, as if the mechanisms are wearing out. We understand how tempting it is to attribute to time itself "the ravages of age." Are they not programmed into the nature of things, into the fundamental princi-

---

[4] The journal *Science* published these results with a commentary entitled "Have 25-Million-Year-Old Bacteria Returned to Life?" *Science* 268 (1995): 977.

ples of physics? The first theories of biological aging were indeed based on the idea of inescapable, almost purely physical deterioration. Today, these theories are outmoded. But let's look at them in their simplest form, to pinpoint an essential aspect of biological aging: namely, the increase in mortality rate with age. To put it another way, the older one is, the greater the probability of dying in the coming year.

Of course, the more we advance in age, the luckier we are to still be alive. This simplistic observation illustrates the upper part of the famous age pyramid. No matter what population you look at, the pyramid pitilessly terminates in a point. The oldest age ranges are the smallest. Is this simply due to the progressive disappearance of individuals, who would statistically all be as likely to be carried off by an illness or accident? Mortality would in that case be constant, regardless of age. This would be the simplest model of wear, which little by little reduces the number of glasses and plates in the cupboard. Suppose you received a set of a dozen champagne flutes for a wedding present, and that each of them has one chance out of ten of breaking during a year. When you celebrate your silver wedding anniversary twenty-five years later, you can mathematically count on having 0.86 flutes. In practice, you should be happy if there is still one intact, for you and your spouse to share!

This same mathematical curve (called an exponential curve; see Figure 2.1) also describes the decay of radioactive atoms. Each atom has the same probability of decaying per unit of time, and this probability of decay is a characteristic of the radioactive element in question. It is usually expressed in terms of a radioactive period, or half-life, which is the time required for half of the atoms to decay. After two periods, only half of half of the original atoms remain, and so on. An atom that is still present after one million years has exactly the same annual probability of decay as its many "brothers" that were there at the first instant. As shown in Figures 2.1 and 2.2, in theory there is no upper limit to the survival of a fraction of the population if it is subject to this type of mortality, independent of age. A practical limit, related to the initial size of the population in question, does exist, however. It corresponds to the time at the end of which the surviving fraction represents only a single individual.[5]

---

[5] This time is on average $T\log_2 N$, where N is the initial number of atoms, and T is their radioactive half-life.

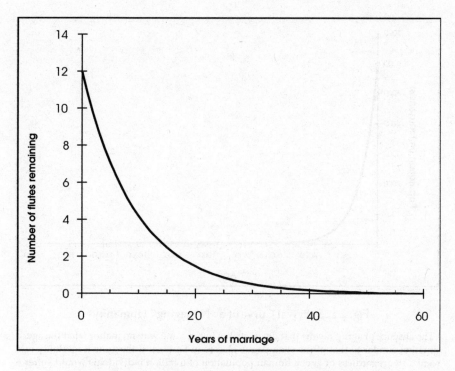

Fig. 2.1.: Survival Curve of a Set of Champagne Flutes*

* The annual probability of breaking a flute depends on the number of parties given during the year and the dexterity of the drinkers (and the person washing the dishes!). It is possible, even if this is not always the case, for these two parameters to remain constant during the life of the couple in question. Here, we assumed that each flute had one chance in ten of being broken each year.

If, as in the case of champagne flutes or radioactive atoms, the mortality rate were constant in the human species, with a "half-life" of around seventy-five years (or roughly the life expectancy in developed countries), one quarter of each age range would reach 150 years of age. There would remain close to one out of a thousand people who reach 750 years of age. Or, if the annual probability of dying was 0.1 percent for the entire course of a human life (the approximate value observed currently in developed countries at thirty years of age), the oldest human on Earth, given a population of six billion inhabitants, could reasonably reach 15,600 years old (see Figure 2.2). The difference between these figures and the harsh reality is striking, since the longest lived human of all time is without a doubt the

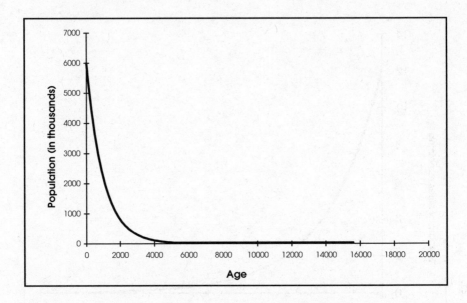

Fig. 2.2.: Survival Curve of a "Nonaging" Humanity*

* The absence of aging means that death strikes in the same way no matter what the age, a situation similar to that of the champagne glasses in figure 1. If the annual mortality rate were 0.1%, regardless of age, a human population of 6 billion individuals would suffer 6 million deaths each year. This implies the same number of births to maintain a stable population. In this case, (population equilibrium) the survival curve represents both the average number out of six million newborns annually who will survive to various ages, and the average number of individuals of various ages living at a given moment. Based on statistical expectancy, the oldest individual would be a little more than 15,600 years old.

French woman Jeanne Calment, who died in August 1997 at the age of 122 years and a few months.

The explanation for this paradox, you have surely understood, is that the mortality rate of adults in fact increases with age.[6] Not only does the age pyramid shrink at the top, but the *rate* of shrinking accelerates. What is surprising is that a relatively simple curve describes this increase. In 1825, Benjamin Gompertz described this relationship for humans—a mathematical formula that is now called a Gompertz curve. Another surprising thing is that the mortality of most species for which data are available complies with a Gompertz curve, generally starting at the age of sex-

---

[6] In humans, this increase occurs probably at the end of adolescence. We will discuss this topic at more length.

ual maturity. Only the values of the two parameters used in the mathematical formula change, as they are characteristic of each species. This brings us to an essential aspect of aging—defined as all of the processes that contribute to the increase in mortality rate with age, for a given species.

No one knows why Gompertz's law is virtually universal. It is totally empirical and is not based on any fundamental principle or gerontological theory. Despite the shortcomings of this model discovered in recent years,[7] it remains one of the pillars of works on aging and is worth closer examination. In its initial form, it was written as m(a) = Mexp(Ga), where m(a) is the mortality rate at age a. If we take the year as the unit of time, m(a) is the fraction of that age population that will die in one year. For a given population, then, this equation means that m(a) is equal to a constant, M, that is characteristic of this population—but independent of age—multiplied by a number that increases exponentially with age. In practice, for every year that goes by, the probability of dying increases by a fixed proportion. This formula is, with less enjoyable results, the same principle that applies in compound interest savings accounts.

Let's compare the survival curves obtained using Gompertz's law with another mathematical model—one that describes mortality that is not constant but increases in "only" a linear fashion (Figure 2.3). To simplify, let's start at an age of thirty years, with an annual mortality rate of 0.1 percent, which corresponds more or less to the reality in developed countries. Between thirty and forty years, this rate goes from 0.1 percent to roughly 0.3 percent. If the increase remained linear, 0.2 percent every year, we would obtain the survival curve drawn in bold (Figure 2.4). We see that the acceleration associated with the exponential factor results in a much more abrupt drop in survival (solid line). This drop-off establishes fairly accurately the maximum longevity, which at that point hardly depends at all on the size of the population.

The American gerontologist Caleb E. Finch[8] suggested describing the variation in mortality rate using two parameters related to the M and G of Gompertz's equation. These two parameters describe the time it takes for

[7] See chapter 8.
[8] Finch is author of an important reference work: *Longevity, Senescence and the Genome* (Chicago: University of Chicago Press, 1990).

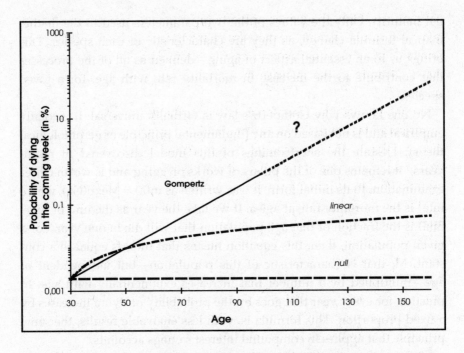

Fig. 2.3.: Different Types of Theories on Aging*

* The most commonly accepted theory of aging is statistical, namely the increase in mortality rate with age. In the absence of aging, mortality remains constant (as shown in figure 2, represented here by a horizontal line). The curve that most closely reflects human reality, proposed in 1825 by the British actuary Benjamin Gompertz, is sadly very different. On this semi-logarithmic scale, it is shown as a line on which the probability of dying is more than 25% per year beyond the age of 100 (this figure indicates the probability of dying per week). If the trend continued, we see that any fortunate beings reaching 160 years of age would have barely one week left to rejoice at still being alive. Recent results suggest that mortality increases less quickly for centarians. In other words, those who are oldest age less quickly! The Gompertz model seems to hold true for a great many species, as does the deceleration of aging at the oldest ages.

the mortality rate to double (for the sake of brevity, we will call it the aging time) and the initial mortality rate. For a given population, the aging time is inversely proportional to the rate of increase in mortality with age. The initial mortality rate represents the "starting point," or the base mortality rate, generally observed around the age of sexual maturity. As shown in Table 2.1, the values vary greatly from one species to another, and we are rather better off than most. Among the species that have a relatively well-

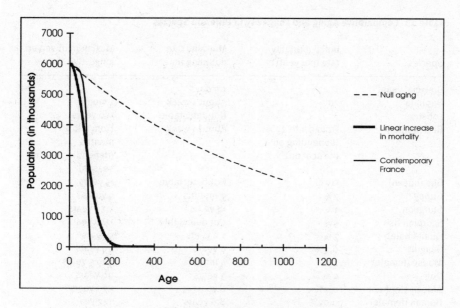

Fig. 2.4.: Different Types of Survival*

* With the same starting parameters as for figure 2, we see here the dramatic effect on survival of mortality that increases exponentially with age. The drop in survival becomes so rapid that maximum longevity is almost independent of the population size. Even if the increase were "only" linear (see figure 3), the oldest individual of an initial group of 6 million individuals would hardly exceed 400 years, compared to 15,000 in the absence of aging.

defined maximum life span, humans do in fact age the least rapidly, with a mortality rate that doubles roughly every eight and a half years.

In extreme cases, the phase of accelerating mortality does not become manifest until the very end of life; however, it may then be so sudden that it is difficult to demonstrate as exponential. Certain bamboo species flourish for 7, 30, 60, or even 120 years (depending on the species) before blooming just once and then quickly dying. Such precipitous decrepitude is often related to reproduction. It is seen in salmon, who return to their place of birth to reproduce, and then die.[9] Because of this suddenness and the specificity of the mechanisms involved, we tend to separate sudden death, directly related to reproduction, from gradual aging, which is the most common case. Finally, it is important to remember that the maxi-

[9] In chapters 3 and 4, we examine in more detail the link between death and reproduction.

Table 2.1. Comparative Aging and Longevity in Different Species

Species	Initial mortality rate (per year)*	Mortality rate doubling time*	Maximum observed longevity
Brewers' yeast	?	1.5 days	?
Daphnia	70%	About 1 week	1 month
Lobster	?	Not detectable	>50 years
Bee (worker)	From 0.1 to 20% (depending on the season)	About 1 week	From 2 to 19 months (depending on the season)
Bee (queen)	<10%	Not detectable	>5 years
Guppy	7%	9 months	5 years
Sturgeon	1%	>8 years	>150 years
Scorpion fish	5%	Not detectable	>140 years
Quail (male)	7%	1.2 years	5 years
Seagull	0.4%	6 years	49 years
Mouse (female)	1%	4 months	4 to 5 years
Dog	2%	3 years	20 years
African elephant	0.2%	8 years	> 70 years
Human (female)	0.02%	8.6 years	122 years

Source: C. Finch, *Longevity, Senescence and the Genome*, University of Chicago Press, 1990.
*For almost all wild species, the margin of error in the first two columns is on the order of a factor of 2.

mum longevity observed in a species does not always give an accurate idea of its aging time. Bamboo age quickly, but only after having lived in excellent health for many years. Bats and humans age at approximately the same rate, yet their maximum longevity is very different because bats have a much higher initial mortality rate. In numerous wild species, few individuals have the time (or, for optimists, the luck) to grow old.

From the Statistical to the Individual; from Mathematical Models to Concrete Mechanisms

Up until now, we have talked about mortality and aging in essentially statistical terms. We have seen that the actual definition of aging, for scientists, is statistical in nature. The Gompertz model uses different parameters for different species. What's more, for a given species, the aging time

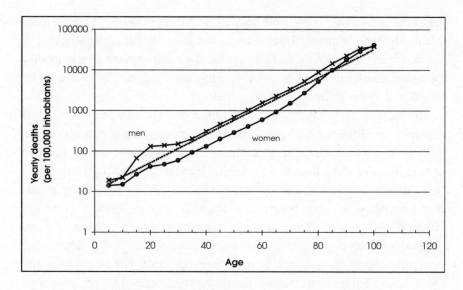

Fig. 2.5.: Mortality according to Age in France (1994)*

* An INED website (http://matisse.ined.fr) provides access to complete data on The Causes of Death in France Year by Year since 1925 (Vallin et Meslé, 1988). We extracted the mortality rates by age group, from 5 to 100 years and over, for the year 1994, separated by gender (women: • symbol, men: x symbol). Excess mortality in men is particularly marked around 20 years of age, but it exists at all ages. On the other hand, between 30 and 70–75 years, the increase with age is pretty much parallel for men and women. Women age just as fast as men! The straight line represents a Gompertzian analysis of the data (for both genders together). It reflects the data with more than 99% accuracy. We can deduce a mortality doubling time of 8.6 years that is identical for women and men.

is largely independent of living conditions. This means that in the human species, aging time differs little from one country to another, whereas absolute mortality rates vary widely. Likewise, the aging time for women is identical to that of men, despite the difference in mortality rates up to advanced ages (Figure 2.5). An even more striking demonstration comes from the sad "experiment" performed on Australian prisoners of war held by the Japanese during World War II. Their mortality was thirty times higher, at an equal age, than that of their compatriots in Australia, but it did not increase faster with age. In other words, subjected to inhumane conditions, they died in greater quantity but did not age any faster. Let us note, however, that in this specific case, the long-term effects, especially on the life expectancies of the survivors, were not studied. Aging, in the statistical sense, was measured only during the period of stress, which, thankfully, was relatively brief. This is not the case for comparisons between different countries. To summarize, the environment affects mostly the initial mortality rate but not the aging time, which can therefore be considered a species-specific characteristic.

Life expectancy is often used as an indicator for comparing different countries or different time periods. Its popularity undoubtedly stems from the fact that it allows us to estimate the average number of years we have to live. For more than two centuries, insurance companies have used "life tables" to calculate life expectancies and thereby determine insurance premiums. In fact, Benjamin Gompertz was an actuary. How are these tables drawn up? For a given age group, at a given moment, we calculate the number of expected survivors each year, if the age-dependent mortality rates in these future years remain the same. For example, in the life tables for babies born in 2002, the mortality rate used for the year 2032 is the mortality rate recorded for the population thirty years of age in 2000 (the most recent year for which data are available). The average number of years of life remaining is calculated by averaging over the entire initial cohort. Remember that this average, which is the life expectancy, is not the same as the median, which is the number of years after which half of the individuals are dead. This difference is due to the asymmetry of the death curve based on age. For example, with 75.3 years of life expectancy at birth for a median death age of approximately 78 years, an U.S. newborn has a little more than 50 percent chance of reaching his or her life expectancy.

These life expectancies are somewhat pessimistic extrapolations in developed countries, where the mortality rate in almost all age ranges has been decreasing—except in time of war—for more than 150 years. If this trend continues, life expectancies based on current values will inevitably be lower than the real figures. In the most privileged human populations, life expectancy at birth has more than doubled over the past two centuries,[10] boosted by the spectacular reduction in infant mortality and the widespread use of vaccines. And the movement is far from over. Currently in France, life expectancy at birth continues to increase by roughly three

---

[10] In contrast, life expectancy probably changed very little from the prehistoric times through the Middle Ages.

months every year, and is now more than seventy-four years for men and eighty-two years for women.

In contrast, maximum longevity has probably not changed for thousands of years, even if it is impossible to be sure in the absence of reliable records. To better pinpoint the progress at the oldest ages, and to evaluate the impact of population aging on healthcare costs, demographers have developed new indicators, globally referred to as health expectancies (e.g., life expectancy without handicaps or major disability). The main goal is to know whether the years gained are mostly morbid, ruined by disease, or rather years with full physical and mental capacities.

Things become complicated when we leave the field of statistics and mathematical models to look at aging on an individual level, to analyze the underlying biological processes. The first difficulty is the definition of aging as an increase of mortality rate with age. It is obviously impossible to measure the mortality rate of an individual organism; an organism can only be alive or dead. To be able to say that it is aging—and, if possible, to quantify its aging—we need a suitable "marker," a measurable parameter for each individual. Biologists record all the types of changes they observe as the individuals of a given species advance in age, and they attempt to establish any correlations between these changes and changes in the mortality rate with age. In practice, the aim is to find, for each species, a "biomarker," or physiological parameter,[11] that can be used to evaluate the probability of a given individual's dying during the coming year. This measure of probability provides an indication of the aging of the individual in question.

Many biological parameters have been studied, of course, including respiratory capacity, blood pressure, skin elasticity, and so forth. At a given age, these measurements can indeed define groups at risk, whose mortality is higher than the average for that age group. Insurers, for example, know that high blood pressure is a risk factor. But these are not universal markers that could conclusively apply to every individual. Further, we can say that high blood pressure (or smoking) kills, but not necessarily that it causes one to age. The most conclusive information for estimating the life expectancy of an individual is still his or her date of birth. To date, no "biomarker" for aging is more reliable than actual chronological age.

---

[11] Any such biomarker would have to exclude any already established disease.

Aging Never Killed Anyone

Does this conclusion mean we have come full circle and should indeed attribute "the irreparable damage of years" simply to the passage of time? No, provided we come up with a more accurate description of this apparent damage, as well as how it relates to the acceleration of mortality. Authors such as Alex Comfort and Bernard Strehler, whose works contributed to gerontology research, did just this in the 1950s and 1960s. Along with Peter Medawar, they defined the aging of the individual as an increase of its vulnerability to all types of aggressions, which strike at random. According to them, no death is totally "natural," since no one dies simply from the weight of years. Weismann had already noted that "one of the most thorny problems of all physiology [is] knowing what is the cause of death." However, he still distinguished between "normal death," caused directly by aging, and death that results from the decreased capacity of the aged organism to "resist harmful external influences of little importance."

The distinction between disease and susceptibility to disease is difficult but essential for several reasons. First, it establishes a hierarchy of causality levels.[12] Today the predominant vision of aging resembles Medawar's. The premise is that aging is not so much a direct cause of mortality as the ensemble of processes that increase the vulnerability of the organism to the direct causes, such as infections, tumors, or occluded or broken blood vessels. One argument supporting this view is the remarkable similarity, which we have already mentioned, of the aging times among different populations of the same species. This similarity contrasts with the variability of not only the rates but especially the causes of mortality. The curves remain parallel, whether the individuals die primarily of cancer, cardiovascular disease, or something else. Similar results have been obtained with lines of genetically homogeneous mice. Each of the lines, due to its specific genetic characteristics, is more likely to develop certain diseases with age. Mortality rates, even though they are shifted, nevertheless have almost the same slopes. We can interpret this parallelism by concluding that all the individuals of a given species, while remaining diverse, still share certain basic aging mechanisms[13] that both increase the

[12] The difficulties involved in biological causality are discussed further in chapter 8.
[13] Recent genetic data suggest the possibility of distinguishing, within a population, subgroups that age at different rates. This finding remains a controversial issue.

organism's vulnerability and act upstream of any pathology. The pathology that actually appears depends on the constitution and the environment of each particular organism, with, of course, plenty of random chance thrown in. Although we can say that diseases that preferentially strike older individuals are "aging-related," aging itself is not a disease.

The distinction between vulnerability due to aging, on one hand, and diseases that are the direct causes of mortality, on the other, contrasts processes that are intrinsic to the organism (and ultimately inescapable) with more external, random factors. This distinction also emphasizes deleterious changes—those that cause the organism's defenses, in the broad sense, to deteriorate with age. Based on this idea, Bernard Strehler came up with four criteria for recognizing those changes in an organism that are truly aging related. The changes must be universal (within a given species), intrinsic, progressive, and deleterious.

The processes that so accurately limit the longevity of a species should indeed show up in all members of that species. This criterion—universality—eliminates most hereditary diseases[14] and the effects of certain environments (which may seem to have something to do with age, when in fact they simply increase with the length of exposure). In contrast, an organism's vulnerability to a particular environment may increase with age, in all individuals, even if it reveals itself only in those who have been exposed. Studying aging means studying the processes that are involved in causing this increased vulnerability. For example, solar radiation can cause cancer. If the incidence of cancer increases with age, it is partly because old individuals are less effective at fighting off the formation of tumors.

The effects of the radiation can also be cumulative, thus contributing to a higher incidence with age. The second criterion proposed by Strehler—intrinsic nature—helps rule out aging as the cause of this higher incidence, which, in this case, would be due to prolonged exposure to a particular environment and would not be intrinsic to the organism. In our example, a potential effect of age should be evaluated at equal exposure lengths—or even better, at equal radiation doses. It is not always easy to know whether a certain "effect of age" meets the second criterion or not. If we can eliminate it by modifying the living conditions of the organism,

[14] Some hereditary diseases partially mimic normal aging, but in rapid motion. They serve as models in the search for genes involved in aging processes (see also chapter 5).

then the answer is probably no. By definition, the basic aging processes must occur even in the most favorable environments.

The progressive nature of aging (the third criterion) facilitates the onset of disease, but contrasts with the sudden nature of pathological manifestations. We can easily conceive of a vulnerability that changes continuously, progressively, whereas a disease is either present or not, even if it may evolve more or less rapidly. The last criterion—the deleterious nature—helps eliminate certain effects of age, such as gray hair or the baldness that affects many males of our species.

So Why Do We Age?

Armed with these criteria, Strehler classified the effects of age according to their origin, and especially according to the level of organization, from populations (which exhibit an increasing mortality rate with age) to molecules. Unfortunately, this effort did not shed any new light on the process of aging, although it did highlight the importance of cellular factors. If a tissue or a vital function is less effective, or less resistant to external aggression, in an aging individual, the explanation lies either in the individual cells that form this tissue (or perform this function) or in the interactions among these cells. Since the cell is the basic unit of the living, this may not seem very surprising. Drawing our attention to the cell reminds us, however, that aging is not simply wear caused by time. In fact, a cell is constantly being renovated. Most of the molecules that compose it today will not be there tomorrow. Likewise, almost all tissues are constantly being renewed. Most of the cells of an organism are much younger than the organism itself. Our red blood cells have a life span of twenty days, the cells of our digestive epithelium are replaced every three days, those of our pulmonary alveoli every week. Only our heart and our neurons are with us from the start. We may be only as old as our arteries, but our arteries can never be as old as we are!

Hence the idea of first examining this minority of cells and molecules that do not renew themselves in the adult organism. This includes, of course, the nerve cells (neurons) that cannot divide (they are said to be postmitotic) and that have almost no precursor cells that could potentially replace them in adults. They can only diminish in number with age, since whenever neurons are destroyed, accidentally or through disease, the damage is irreversible. Even though on the whole, the intellectual capacities of elderly people in good health are remarkably well preserved, the incidence of neurological pathologies, such as Alzheimer's disease, seems to increase exponentially after the age of sixty. Are these diseases due to the normal aging of the brain? The question is still being debated, but recent studies[15] suggest—fortunately—that these are in fact age-related diseases and not an inevitable aspect of aging; a large proportion of very old people may never experience them. The heart and most of the skeletal muscles are composed of postmitotic cells, and they have little or no ability to regenerate. Still, aging is not just a problem affecting the brain, heart, and muscles. One of the major gerontological observations is that aging, in most species, is not associated specifically with the deterioration of a particular organ. We cannot say that the factor limiting longevity is simply a single physiological function.[16]

Other biologists believe that we should look at intercellular reactions and deregulation rather than at a certain type of isolated cell. They emphasize a balance of hormones, or balance within the immune system. Clearly, many cell lines have a replicative potential that far surpasses the life expectancy of the individual. Various experimental manipulations can immortalize—this is the technical term—almost any type of cell. As Weismann said long ago, there is no absolute biological necessity that limits the longevity of any cell line. But the direction of causality is often difficult to establish, since the deterioration of a cell as the organism ages may very well be the result of aging rather than the (or a) cause. The relationship between, on one hand, cell senescence, programmed cell death, or any other form of attack on the functional integrity of the cell, and, on the other hand, the natural death of the organism is likely to remain a subject of debate for quite some time.

Faced with such a tangled web of causes and effects, with the difficulty of defining the very nature of aging, we would be in dire straits if we didn't have a conceptual framework that takes into account the extreme

---

[15] See chapter 5.

[16] Chapter 4 demonstrates the theoretical scope of this observation. The exceptions, such as that of Metchnikoff's silkworm moth discussed in chapter 1, are trivial. Incapable of feeding themselves in adulthood, they are obviously not meant to last unless force-fed by breeders!

prevalence of natural death. This framework does exist and is discussed in later chapters. It suggests an unexpected and paradoxical answer to the question "Why do we die?" It also offers a valuable guide to research into the "how?"—that is to say, research that seeks to elucidate the various death-inducing molecular mechanisms. This research may one day result in radically new therapeutic or preventive measures. Before presenting the research, however, we first look at the diverse modes of aging and death that exist throughout the living world. Using numerous examples, we then attempt to answer the first question—namely, "Why?"

CHAPTER THREE

AGING AND LONGEVITY ACROSS THE SPECIES

The question is not so much why we die as why we live as long as we do.

George Sacher

The preceding chapter, by insisting on the universality of aging and death, could cause one to think that there is no reason to seek a specific explanation for these processes. Why not raise the Gompertz equation, which seems to describe the age-related increase in mortality so well, to the rank of a fundamental law of biology and leave it at that? The existence of such a law would provide an additional reason to view "natural" death as a fundamental property of life. Linnaeus and his students in the eighteenth century believed that the balance of nature required certain organisms to be eliminated as others came into the world. And it's a sure bet that these naturalists would have been won over by the precision of the mathematical formula proposed by Gompertz a century later. We cannot just leave it at that, however, because the modes of elimination by "natural death" are incredibly diverse, as are the modes of reproduction. How good would our understanding of the living world be if we left out the origins of this diversity? In fact, surrounded by so many means of dying naturally, only one thing can be said with certainty: You only die once. The goal of this chapter is to give an overview of the full range of longevities, ways of aging, and ways of dying.

It is convenient to distinguish three major categories, depending on the intensity and rapidity of aging. Like all methods for classification, ours has an arbitrary element, but it also has the advantage of highlighting some of the crucial questions that the explanatory theories presented in the following chapters must answer. The first category, which we call the "sudden death" model, contains the species for whom death seems directly programmed into either the organism's behavior or its structure. These species generally reproduce during only one season, sometimes laying only one set of eggs (or disseminating seeds only once). More or less sudden death follows immediately thereafter. The second category, termed "gradual aging," contains the majority of species whose mortalities increase with age in a continuous and progressive fashion, and almost always fit the Gompertz model. This category includes organisms whose mortality-doubling time ranges from one or two days (yeast) to several months (certain mollusks, fish, and rodents) to several years (many birds, most mammals). The latter group is where our common notion of aging comes from, and naturally this notion applies best to them. The third and last category, which we will call "negligible aging," is perhaps the most fascinating, both from a human and scientific point of view. It includes species that exhibit negligible aging and indefinite longevity. Even though no definitive proof is possible in terms of eternity, these species seem to open the door to potential immortality.

Sudden Death: When It Pays to Bet It All

Many animal and plant species offer spectacular examples of death following immediately on the heels of reproduction. The final moments often occur so rapidly that they cannot even be described by the Gompertz equation. These examples contributed greatly to the very attractive idea of individual sacrifice for the perpetuation of the species. Certain spiders are emblematic in this respect. In a few species, the female almost always devours the male during copulation. Note that despite the folklore attached to its name, the famous black widow spider (*Latrodectus mactans* to its close friends) does not engage in erotic cannibalism. Its widowhood is nonetheless very real, since the life span of the males is on average ten times shorter than that of females! In many arachnid species, the males do not survive their first, very exhausting, embrace. The same goes for male bees. At the end of their frenetic mating flight, they fall to the ground, spent, often mortally wounded. Their sperm is used by the queen bees for years to come to produce the hordes of worker bees needed to sustain the hive. The grand prize for longevity, in the "adult insect" category goes to a queen ant who reportedly lived for at least sixteen years. No significant manifestation of aging is observed in these royal females, who seem to belong to the third category of our classification system. They might live even longer if they were not killed by their own daughters once their sperm bank, filled only once during the mating flight, runs dry. The daughters, in contrast, seem to suffer gradual aging, thus belonging to our second category, but the mortality data are too scarce to be reliable.

Several species of worms present a maternal form of direct sacrifice on the altar of reproduction. The body of the mother is pierced by its progeny when they hatch. In marine worms, the egg-laying act itself is explosive: When the eggs reach maturity, they blow up the parent's body. Stranger yet, in the beetle *Micromalthus debilis*, the male embryos produced by parthenogenesis (without fertilization) develop in the ovaries. They then leave the mother's body, attach themselves to it, and feed themselves by cannibalizing it.

In these extreme cases, the sacrifice of at least one of the two parents is closely linked to the reproductive process itself. The link between death and reproduction may be less immediate—for example, when the sexually mature adult is simply not adapted for prolonged survival. This situation is encountered in many species of insects, such as the silkworm moth. Metchnikoff was particularly interested in this species as an example of perfectly "natural" death because it is programmed into the nature of things. Usually death in this type of insect is due to the total absence of mouthparts or digestive organs. The latter handicap may be caused by active self-mutilation (destruction of the intestinal epithelium, which is digested from the inside by phagocytosis in the last pre-adult stages of the mayflies). The resulting aphagic adults (incapable of feeding) live off the nutritional reserves acquired during their larval or nymph stages. In many species, these stages are much longer than their subsequent lives, which led the gerontologist Caleb Finch to include all the phases of development, starting from the fertilized egg, in the longevity tally. Some mayflies (whose airborne life span varies from a few minutes to several weeks, depending on the species) have already lived a number of years in the larval, naiad, or nymph state. Perhaps there is no need to pity them for their brief adult existence. Other species are even worse off. Consider the adult female *Orygia* moth, which doesn't even have wings!

The males come and fertilize them in their cocoons, or in some cases, their eggs develop by parthenogenesis.

Sudden death can still be compatible with a respectable old age. This occurs in bamboo of the genus *Phyllostachys*, which also present another characteristic of our first category of natural death: synchrony within a species or a population. Most of the species of this exotic grass have their own particular blooming pattern, from every seven years, to every 120 years or more. The plant, sometimes living for more than one hundred years, dies immediately after blooming. In Japan, the longest cycles have been followed for nearly one thousand years. The cycles repeat with impressive accuracy. All the individuals of a given species bloom the same year, in such unison that it is sometimes called an epidemic. In the late 1960s, *Phyllostachys bambusoides* that had been transplanted to England, the United States, and Russia almost simultaneously completed their 120-year cycle, blooming in synchrony with their fellow plants that had remained in Japan! The periodical cicadas of North America, of the genus *Magicicada*, are another well-studied example of long but precisely measured existence. These insects emerge from their subterranean regions by the millions, at the end of their thirteenth or seventeenth springtime (depending on the variety). Although they then live no more than a few weeks, they could very well challenge the queen bee or the queen ant for the title of the oldest insect on earth. No one knows what biological clocks underpin the multiannual cycles of bamboo or periodical cicadas.

We see in all these species the sudden and almost synchronized death of entire populations, with massive stereotypical changes in all of the individuals involved. In insects, these changes reflect mostly the exhaustion of nutritional reserves, for those that are aphagic, or the wearing out of the organism, in which almost none of the cells renew themselves.[1] The important point is that there is no escaping this massive decrepitude programmed into the very structure of the individual. Death is clearly mandatory. In other cases, the situation is more complex. Certain species—at the top of the list are salmon species and the marsupial mouse—seem to have opted for a reproductive strategy that kills them,

---

[1] This does not rule out prolonged survival (more than one year for female aphagic beetles of the genus *Pleocoma*, who live off the reserves provided by their adult size of several centimeters) or gompertzian aging. Some flies, including the famous fruit fly can also be included in our second category, as can the nematode worm, *C. elegans*.

whereas their constitution would have allowed them to survive much longer. Death in this case appears to be one option among several.

In Australia, the males of several species of small insectivorous marsupials, commonly called marsupial mice, live an intense but brief reproductive period, like that of mayflies. When the mating season arrives, at the end of the austral winter, their sexual hyperactivity (in the laboratory, they can spend up to twelve consecutive hours copulating) leads them to death within two to three weeks, with a rapid, stereotypical deterioration of their general state of health. This result is doubtless due to sudden increased production of corticosteroid hormones by the adrenal glands, which increase in weight by 50 percent. The animals lose weight, stop cleaning themselves, are immunodepressed, and become more and more aggressive. They experience digestive disorders, often developing ulcers that bleed and cause anemia. This degeneration is genuinely due to sexual activity, because most if not all of the changes are prevented by castration. These changes do occur in the absence of females, but they are much less drastic. Males that have not mated can live up to three years in captivity, or more than three times their normal life span in nature, where no marsupial mouse has ever met its father. Just before the birth of the first litters in early spring, the population is composed entirely of females.

Death, which is in this case directly related to reproduction, seems to have an optional component, as demonstrated by virgin males, the survival of females (although few of them have second litters the following year), as well as the existence of very similar species in which the males commonly live nearly three years in the wild. This feature does suggest that there is no strong internal need for this type of organism to experience sudden death, since an only slightly different "version" has managed to escape it. Migratory fish such as salmon and eels show the same thing. Salmon, born in fresh water, live in the sea for several years. They then swim back up the river where they were born, often a Herculean effort, to reproduce. Five species of Pacific salmon (genus *Oncorhynchus*) die immediately thereafter, with a clinical picture similar to that of the marsupial mice.[2] In this case as well, castration has a protective effect, virtually doubling the life span of the fish. In contrast, three related salmon species can recover and make the trip a second or even third time, al-

[2] However, in this case, males and females alike are affected.

though they too struggle during migration. One of the luckiest species, *Oncorhynchus mykiss*, a purely freshwater fish commonly known as the rainbow trout, lives longer and reproduces several times. But a hydrocortisone implant, even in the youngest individuals, provokes the same changes experienced by their ocean-going cousins during the long migrations that hasten their end.

Studies on eels show similar though less certain results, since they travel in the opposite direction: no one knows exactly what happens in the spawning ground of European eels, in the middle of the Sargasso Sea. It is clear in any case that eels live much longer if they are prevented from going back to their original sea.[3] Finally, although almost all fish species that experience sudden death are migratory, not only do not all migratory fish experience sudden death, but the sturgeon, which travels almost as far as the salmon to reproduce, holds the record as the longest lived fish of all!

What conclusions can we draw from these examples of species that experience sudden death? First, that reproduction inevitably kills, either because the sexually mature forms are not designed to last (aphagic insects) or because reproduction is accompanied, almost immediately (e.g. in cannibalism or matricidal birth) or within a short time (e.g. in bamboo, salmon, and marsupial mice) by mortiferous processes. But we can also see the limits of our typology of natural death in what is clearly emerging as a continuum of diversity. The same species of social insects can fall within all three categories, depending on the caste in question: sudden death for the males, rapid but progressive aging for the workers, imperceptible aging for the queens. The workers themselves could also be relegated to the first category, to emphasize the fact that their cells are not renewable. And when the link between death and reproduction is stretched at bit, we often find a sort of transition zone where the link is optional, as illustrated by the different species of salmon.[4]

Lobelia, a herbaceous plant that grows on Mount Kenya, provides a classic example of this transition zone, which has a very literal meaning for botanists. The two species, *Lobelia telekii* and *Lobelia keniensis*, both take forty to sixty years to reach sexual maturity. After their first blooming,

---

[3] According to a Swedish museum, the record life span for an eel is eighty-eight years, or six times longer than the norm.

[4] Taxonomic revisions within the Salmonidae family in the late 1980s blurred the lines of this transition zone even more.

however, their fate differs greatly. *L. telekii* dies almost immediately, whereas *L. keniensis* survives and blooms approximately every ten years. The dividing line between them, at a given altitude, is drawn out by climatic conditions: the first lives on arid slopes, the second in moist valleys. They coexist in the intermediate zones. To understand what determines this division, we must look at evolutionary models, which we will discuss in later chapters.

But let's take a look at the problem now in the terms used by L. C. Cole in 1954. Cole established the basic distinction between the terms semelparous (individuals have only one reproductive cycle) and iteroparous (individuals have several reproductive cycles, regardless of the duration of the cycles).[5] Here is a thought-provoking version of what is sometimes called Cole's paradox: "But why in the world are there iteroparous species, since many other species do just fine with a single reproductive cycle, which is enough to ensure their perennity?" Luckily for the human species and a few others that we discuss later in this chapter, this paradox can be resolved. It is interesting, though, in that it reverses the usual question. In other words, according to a phrase attributed to George Sacher, the question is not so much why most organisms are mortal, but rather why they live so long. For iteroparous species, this often means examining the speed at which they age.

Gradual Aging

In the first category of natural death that we discussed above—the "sudden death" model—organisms develop until they reach sexual maturity, then die relatively quickly, in their prime and without aging, as it were. Indeed, their death is so sudden—it follows so closely on the heels of reproduction—that the concept of age and aging become almost irrelevant; aging, by definition, means running a greater risk of dying each day. Unless we could consider the sudden death in marsupial mice or salmon as a form of extremely accelerated aging. Or (and this boils down to the same thing) we could consider gradual aging as sudden death in slow

---

[5] In general, there is one reproductive cycle per year, but as we have seen, the cycle of certain bamboos lasts 120 years.

motion. Since sudden death is so compacted, it could then be a practical model for understanding aging in the usual sense of the word.

In fact, this is not so, even though in species with gradual aging, the rare individuals that reach the oldest ages do always seem confronted with a sudden death situation; they all disappear within a relatively short period due to the exponential nature of the Gompertz equation. However, even these "survivors" do not all die of the same causes, and the actual causes of death are often difficult to establish.[6] Most of the organs generally remain functional until the very end and would have been able to continue for quite a while thereafter. This variability contrasts with the stereotype of sudden death. In addition, at the oldest of the old ages, mortality may eventually reach a plateau, a very high limit that obviously does not fit the Gompertz equation, and even less the "sudden death" model. The absence of synchronicity is another trait of gradual aging that contrasts with sudden death. Such asynchrony cannot be attributed simply to the diversity of populations, since even within a pure, genetically identical line, death occurs at different times.

Still, despite these essential differences, species that experience gradual aging cover almost the same longevity spectrum as those that experience sudden death, although none has such a short life span as certain mayflies that live only a few minutes. Baker's yeast comes the closest, with a life span of about a week and a mortality rate that doubles in less than two days. Among species that experience sudden death, the longest lived bamboo can rival humans and sturgeons, which are species that age gradually. So a species' life span does not indicate how it ages. To find out how a species does age, we need to know how mortality changes with age, or in other words, we need to have the equivalent of actuarial life tables for the species.

We must insert a brief parenthetical note here, in order to emphasize the limits of any comparative study on aging and longevity, and therefore any conclusions drawn from such a study. We have seen that aging is a phenomenon intrinsic to the organism and affects it even in an optimal environment. To study this phenomenon, a gilded cage could do the trick, but

---

[6] That the actual cause of death in old people is not known is often concealed by the statement "cardiac arrest" on death certificates. For an exhaustive analysis, see Anne Fagot-Largeault, *Les Causes de la Mort, histoire naturelle et facteurs de risque* (Paris: Vrin, 1989).

it would take a great many animals to obtain statistically significant data, especially at the oldest ages. This task is not too difficult when it comes to worms, flies, or yeast, but becomes prohibitive for elephants or dolphins. The conditions of captivity, however ideal they may be, may also distort the results. Field studies, which are more costly, also present at least as many interpretation difficulties. The animals that are captured or killed by humans are often in the worst health. Inversely, those that survive the longest in their natural habitat may form a skewed sample with respect to the entire species. A high extrinsic mortality rate (due to predators, for example) may mask aging, especially since the aging process generally does not manifest itself until some time after sexual maturity.[7] It is important to keep all these issues in mind when interpreting our results, not to mention the generalizations that we draw from the results. Even if these results were all perfectly reliable, they would reflect only a tiny minority of living species.

Let's go back to the Pacific salmon and the *Lobelia* on Mount Kenya. They illustrate the continuity, but also the elasticity, of the ways organisms die naturally. During evolution, similar species acquired very different mortality parameters, which seem to be correlated to their living conditions. This is an important avenue to explore when attempting to answer the question raised by Sacher: Why do some species live so long, whereas the bare minimum—death immediately after reproduction—suffices for so many others? Species with gradual aging provide valuable information because they offer interesting examples of elasticity in longevity and speed of aging.

Some species of nematodes, a class of cylindrical but unsegmented worms, exhibit a life expectancy that ranges from less than one month to more than ten years. One of the nematodes, *Caenorhabditis elegans,* is a favorite among biologists. Its adult life expectancy, only a few weeks, is one of the shortest. The life span of other nematodes living freely in the earth barely exceeds five months, but parasitic nematodes are completely different. Young mermithid worms, parasites of grasshopper nymphs, feed hungrily off their hosts for three months. When the adult worms leave, their pharynx and anus disconnect from the intestine, and they become

[7] As we have seen, having separate data for males and females is essential.

aphagic. In the laboratory, nonfertilized adult females can nonetheless live for more than two years without exhausting their food stores. Like *Pleocoma* beetles, they demonstrate that aphagic does not necessarily mean short-lived. The life span of other parasitic non-aphagic nematodes seems to be proportionate to that of their hosts, where they spend their entire adult life. *Ancylostoma*[8] and filiarial nematodes live for so long in their human hosts that these parasites could be placed in the third category, characterized by negligible aging. For obvious ethical reasons, however, we do not have data on their natural mortality rate.

Many species of insects show gradual aging—so gradual in fact that, in the absence of mortality data, coleoptera such as *Tribolium castaneum* (flour beetle) and *Macrochynus glabratus*, with a life span of two and nine years respectively, can join the social insect queens in the "negligible aging" category. Cockroaches, although they are far behind, are also among the longest-lived insects; certain cockroach species can live up to fifteen months. At the other end of the spectrum, there are the flies and most diptera (e.g. worker bees and ants, many butterflies), whose aging process, though gradual, is nonetheless rapid, with mortality-doubling rates on the order of one week or less. The distinction between gradual aging and sudden death is very fine, especially if we extend the life span continuum to include mayflies. In fact, the classification proposed by Caleb Finch places yeast, *Drosophila*, and nematodes in the same category as mayflies and salmon—namely, rapid aging and sudden death. In the end, our categories are not so very different from his, but we wanted to insist on the nature of the link between reproduction and death (unavoidable or not), as well as on the ambiguities of the classification. In addition, our presentation does not support the idea that sudden death is simply accelerated aging.

Most observations of gradual aging are made on vertebrates. In part due to practical reasons (ease of breeding, accessibility in the natural environment, etc.), such observations are not evenly distributed among taxonomic groups. For amphibians and reptiles, the data are very scarce, indicating above all the absence of sudden death. Gradual aging has been measured in some species of lizards and alligators, but most of the species

---

[8] Parasite of the small intestine that causes anemia.

studied could belong to the category of negligible aging. Fish, which form the third major class of poikilothermic vertebrates[9] have been studied slightly more. Fish catches constitute an abundant sampling pool, but are obviously selected more for culinary reasons than scientific ones. The cost of raising fish is relatively low, if we limit ourselves to small species with few dietary requirements. Fish cover a wide range of life spans—from one year to more than a century—and rates of aging. There are fish species that fall into each of our three categories. While salmon and eels undergo sudden death, in many other species aging is, on the contrary, undetectable. Most fish continue to grow after they reach sexual maturity, albeit more slowly. They are much sought after by fishermen looking to break records and by gerontologists looking for chronological markers. The scales and other mineral parts of the body are in fact the equivalent of annual growth rings on trees. This made it possible to confirm the longevity record for a fish, held by a Lake Ontario sturgeon, at 152 years old and nearly five meters long.

Fish, like all organisms that do not maintain a constant body temperature, are highly affected by the ambient temperature. Within the limits of viability, the lower the temperature, the more slowly the organism functions, and the more slowly it ages. This concept has been illustrated in flies and nematodes, which live twice as long when the temperature is lowered by ten degrees centigrade, and it seems to hold true for fish as well. One of the most spectacular examples involves the brook trout, introduced into a lake in the Sierra Nevadas that was colder and less nutritive than its waters of origin. Its life span quadrupled, going from barely six to more than twenty-four years, and sexual maturity was delayed, occurring only after about fifteen years. In mammals, calorie restriction, that is to say, a low-calorie diet, results in a similar general slowdown of development and aging. We will come back to the implications of these results. In any case, they corroborate the notion of elastic longevity mentioned earlier.

Incontestably, we have the most data on aging and longevity in birds and mammals, though here again data are available for only a small minority of species, at most 5 to 10 percent. This relative abundance has

---

[9] More commonly known as "cold-blooded" as opposed to "warm-blooded" animals—birds and mammals—which are homeothermic.

given rise to numerous comparative studies as well as a few hypotheses on the mechanisms that determine longevity for each species. It is particularly valuable to study mammals since their postnatal evolution, aging included, seems to follow remarkably similar paths from rodents to humans, though at very different speeds. Gompertz studied this class of vertebrates for reasons that had little to do with biology, but his law was subsequently shown to apply most accurately and broadly to mammals.

Another universal characteristic of aging in mammals, at least for females, is the progressive loss or degeneration of their eggs. On average, the eggs are lost completely when the females reach two thirds of their life span. Under protected conditions, then, practically all species of mammals experience the equivalent of menopause during the last third of their lifetime. This is not an artifact of captivity, because cases of post-menopausal females have been observed in the wild, not only in apes but also in whales, elephants, opossums, and others.[10] This is a rare if not unique example of massive loss of an entire category of adult cells, which affects all the species of a taxonomic group. In contrast, the disappearance of neurons usually affects only certain regions of the brain, resulting in the loss of only a small portion by the end of normal life. Other functional deficiencies, such as decreased immune response, presbyopia, and osteoporosis, almost always appear during the second third of life, regardless of the life span of the mammal in question. These similarities highlight the value of research on aging in nonhuman mammals.

These similarities also evoke that old notion of "natural" death programmed into the very nature of the living. Indeed, since all mammals endure comparable forms of decrepitude, can we not conclude that this process is part of the nature of mammals, just like, say, nursing? There is no definitive response, but the example of mammalian dentition shows, once again, the truth about "natural" limits that appear to be insurmountable. In the wild, most mammals die before their teeth become so worn as to pose a crucial problem. Having teeth that wear down does, however, place an upper limit on the animals' life span, unless they have access to dentures or a soft diet, as in the case of pets and humans. Two groups of mammals, elephants and manatees, have found a way to preserve func-

---

[10] However, the total, and rather sudden, cessation of ovulation seems to occur almost exclusively in primates, although probably also in whales.

tional teeth for longer, without human assistance.[11] Their respective diets put their teeth severely to the test. Elephants have only one series of teeth, few in number, that erupt gradually. The first premolar comes in before birth, while the three molars on each jaw quadrant erupt successively at a rate of one every ten years. Tooth replacement is practically continuous. In addition, each tooth grows almost horizontally, in layers derived from the same tooth germ. These layers bind to the back of the tooth, which wears away from the front. The new tooth continues to push the old one from behind, until it falls out. Human dentition also has delayed eruption. The most striking example of this is wisdom teeth, but all of the adult molars (also three per jaw quadrant, counting the wisdom teeth) could be just very late milk teeth. Elephants, however, have taken the concept much farther and have added unique features. This adaptation is essential for their longevity, which, at more than seventy years, is second among mammals only to our own.[12]

Manatees are aquatic herbivores sometimes called sea cows that live at the mouths of tropical rivers. Their marine diet is mixed with large quantities of sand and is very abrasive. They don't have just six or seven teeth per jaw, but thirty or more, without apparent limit. These teeth seem to be separate dentitions, although they are replaced horizontally, from the back of the jaw. The manatees' tooth replacement plays an important role in their life span, which, through less impressive than that of elephants, can still top thirty years.

At this point, there is already at least one lesson to be learned from the diversity and flexibility of life histories from one species to another: the absence of a major constant. No trace of absolute necessity appears to require living beings to disappear after a certain number of reproductive cycles. No fate appears to condemn this or that class of organisms to die off after a certain age limit beyond which they cannot live. On the contrary, when it seems necessary or useful (the next chapter will attempt to clarify what we mean by this), a species always seems capable of putting the inevitable off until later.

---

[11] Incidentally, it is possible that manatees, which supposedly engendered mermaid legends despite their two hundred kilos, are evolutionarily closer to elephants than to whales or seals.

[12] We could almost say that elephants have adapted *to* their longevity. The next chapter will show that this not just a joke.

General Laws for Gradual Aging?

Isolated comparisons have helped us highlight the role of an ecological or behavioral factor in aging and longevity in extreme cases. Let's go back now to gradual aging, not in extreme cases, but in general. Given the relatively large amount of data available for mammals and birds, it was tempting to extend the comparison to all the species of these two classes to identify some general—and, if possible, quantitative—laws similar to the Gompertz equation. Perhaps we could find clues to the processes that determine aging, just as physicists and chemists used Mendeleev's periodic table to design and test organizational models of inorganic matter. The first general "law" proposed in the nineteenth century emphasized the link between the size, or more accurately, the mass, of an organism and its life span. It seems relatively obvious that on this scale, a mouse's few grams cannot measure up against a whale's several tons. For a whale calf, reaching adult size may be literally but child's play, but it does seem a bit harder than for the baby mouse. This is really just a simple, general rule of proportionality. One way to look at it is to say that the larger the adult size of an organism, the longer it has to wait to reach it. This rule is also true of the giant redwood, compared to a modest shrub.

Can we go further and derive a quantitative relationship between longevity and structural or physiological characteristics of mammals and birds?[13] Many authors have made the attempt, in the form of so-called "allometric" equations. J. Huxley coined this term in biology around 1930 to designate non-linear relationships. The general form of the equations proposed by Huxley is $Y = aX^b$.[14] The best known of these equations describe the relative variations of different organs or parts of the body, from one species to another. These variations are subjected to considerable structural constraints, and can be expressed in almost mechanical equations from which the organism can hardly physically deviate. For example, muscle mass must grow at about the same rate as the total mass that it is required to move. On the other hand, bone mass must grow more quickly. A simple calculation shows why. Suppose that the size of an organism is

[13] The parameters taken from Gompertz's equation (aging time, initial mortality rate) would be preferable to life span, but they are known in many fewer species. The relationship between the mortality doubling rate and the mass of the organism nonetheless seems fairly similar to the one described below for longevity.

[14] If b = 1, we have the special case of a linear relationship.

to be doubled, in all three dimensions. Its volume, and therefore its mass, will be multiplied by eight. To support it, the area of bone in cross-section, which determines bone resistance, must increase by the same amount. But the weight of a bone is roughly proportional to its volume, equal to its cross-section times its length. Since our hypothetical animal is two times bigger than before, so must be the length of all of its bones. Its skeleton is thus not eight, but sixteen (eight for bone cross-section times two for bone length) times heavier! The coefficient b that relates bone mass to total mass is 4/3, and the relationship is called hyperallometric. This more rapid increase in bone mass imposes an insurmountable limit on the size of any animal that must support (not to mention move) its own weight, namely the limit at which bone mass catches up with total mass. Archimedes' principle allows whales to easily free themselves of this constraint, as Galileo realized more than three centuries ago.

The range of mammal life spans, from roughly one to one hundred years, cannot even compare to the range of masses, from one to four million grams. Unlike bone mass, longevity increases much less quickly than the mass of the organism. The relationship is called hypoallometric, with a "b" coefficient close to 0.2 and relatively similar for mammals and birds. Still, there are considerable deviations from the mathematical average obtained using this formula. In statistical terms, this means that the differences in mass among species do not account for differences in life span in half of all cases. The mole rat, for example, which weighs only a few dozen grams, lives for ten years, twice as long as the much larger "normal" rat. Likewise, a finch weighing twenty-five grams can live for twenty-nine years, much longer than the five- to twelve-year life span of most Gallinaceae, some of which weigh several kilograms. These extreme examples show that longevity cannot be determined by the mass of the organism, the way bone mass or organ size can. Though similar in form, the corresponding relationships are of a very different nature. In one case, the equations link body features whose interdependent relationship is often obvious, allowing us to predict the form of the equation. The two terms have the same physical dimensions. In the other case, the two variables do not have the same physical dimensions (years versus kilograms), and the relationship expressed by the equation is merely numerical. These are correlations, and weak ones at that, between measurements for which nothing indicates, a priori, that they should be causally related. We will return to this fundamental distinction between correlation and causality.

To improve the correlation, we can try to include more parameters. For mammals, for example, we would have an equation to represent life span in years: $0.66B^{0.6} \times M^{-0.4} \times O^{-0.5} \times T^{0.25}$, where B is the brain mass, M is the total mass (both in grams) O is the resting oxygen consumption rate per gram per minute, and T is the body temperature (in degrees Celsius). Out of eighty-five species, we account for more than 80 percent of the range of life spans, compared with only 50 percent when mass only is considered. How do we interpret this new equation? First, body temperature has a small influence, since its exponent (0.25) is the lowest in absolute value. Body temperature varies very little from one mammal species to another. What is interesting is that longevity no longer varies in the same direction as the mass (M) of the organism, since its exponent is negative (minus 0.4 instead of 0.2). This apparent contradiction can be explained relatively easily: on average—but only on average—the brain mass (B) is proportional to total mass. Since the exponents are added together, if you systematically replace the specific value of C with the average value as a function of the mass, you obtain an exponent of 0.6–0.4, or 0.2. However, if you take into account the differences that may exist between two species of equal mass, you will see that the prize for longevity goes to the biggest brain!

Human encephalization and vanity being what they are, this conclusion seems self-evident. The most intelligent species live the longest. Let's assume for a moment that this is true. Is it because smarter species age less quickly? To find out, we would need equations that handle the initial mortality rate and the mortality-doubling rate separately. In the absence of accurate data, Finch observed that the first parameter is much more sensitive to the environment than is the second. And increased cognitive and motor capacities, due to a more developed brain, would seem to enable a species to better control any given environment. We would therefore expect a big brain to increase life span by reducing the initial mortality rate (for example, by helping fight off predators) rather than by slowing down aging. Finch saw this behavior as a significant contribution to the increase in human longevity over the last few hundred thousand years of evolution of our species. It is true that the correlation between relative brain size and longevity is very marked in the primate order. However, the general significance of this fact is doubtful. In fact, in mammals as a whole, the relative size of the spleen or the liver is just as well, if not better, correlated to longevity! So much for our vanity.

The last parameter of our equation—oxygen consumption per unit of mass and time—has also received much attention. Also known as basal metabolism, it reflects the energy expenditure that the body at rest requires to function. The idea that a limited energy supply to the tissues could influence longevity is not new. Weismann, for one, mentioned this possibility. But he backed away almost immediately, noting that birds have a more intense basal metabolism and burn three times more calories during their lifetime, overall, than do mammals. There is no general limit to the "work" that a gram of living tissue can supply.[15] The idea nonetheless survived until it was consecrated by R. Pearl, in the 1920s, under the name of the "rate-of-living" theory. The brook trout of the Sierra Nevadas, which we considered earlier, offers one possible example of this theory. Likewise, the lengthy life span of bats has been attributed to their long winter torpor, during which their metabolism slows considerably. But there are also counterexamples. Tropical bats, which do not hibernate and seem to burn ten times more calories per year than their cousins, still live just as long. For this and other reasons, Pearl's theory is not tenable, at least in its original form. It did, however, inspire many fruitful approaches. One of the most popular current hypotheses looks not at the rate at which calories are burned but at the damage caused by burning calories, specifically by the production of dangerous free radicals.[16] The mechanisms that reduce or eliminate free radicals, or that combat their effects, may be the true factors that limit longevity.

Informative Comparisons

At this point, readers may well be asking themselves whether all these correlations are useful at all, given how inaccurate and difficult they are to interpret. The answer is yes (otherwise, why spend so much time and mathematical effort on them?), provided they are interpreted as trend guidelines, or reference points. If certain groups of species do not conform to a general trend, or if the trends manifested by two groups of species differ, this result can be seen only if these trends and correlations have been

---

[15] Weismann's metaphoric expression is worth quoting here: "One must not think of the organism as a pile of fuel that disintegrates more quickly into ashes the smaller it is and the faster it burns, but as a fire to which one can always add more logs, and that can be kept burning as long as necessary regardless of whether it is burning fast or slow" ("The Duration of Life," op. cit.).

[16] See chapter 5.

established in the first place. They are the standards by which deviations can be measured, and whose causes we can go on to seek. We saw, for example, that the life spans of birds and mammals vary roughly according to mass. On a graph, this is shown as two parallel, but not superimposed, lines. On average, for a given mass, an avian species can expect to live 70 percent longer than a mammal species. This advantage is less marked in birds that are the least capable of flying, such as penguins or gallinaceans. Thus, the life span of the wild turkey or the pheasant is approximately two times shorter than would be expected based on its weight. In contrast, finches and hummingbirds, which are known to be excellent fliers, deviate from the allometric prediction in the other direction.

The advantage of an airborne lifestyle is also seen in bats. At equal weight, they live three to five times longer than rodents, sometimes for more than thirty years. Even just gliding, as some squirrels (those known as flying squirrels) do, seems to help. The life spans of a dozen or so species of this type, which belong to several different evolutionary lines, have been measured and shown to be an average of 70 percent longer than that predicted solely by the correlation between life span and mass in mammals. The longevity of these gliding animals is not necessarily related to their aerial abilities. But this collection of results does suggest the existence of a link between taking wing and survival. The most commonly held hypothesis is the protection offered by the vast skies; the mortality rate seems lower for these animals. Another similar argument is that one third of the bird species that have become extinct in recent centuries had lost their ability to fly. However, these "degenerated" birds represent only one percent of all avian species. These observations may seem trivial, in that they fit in with the very general assertion that the greater our chances of escaping our predators, the longer we are likely to live—and fleeing to the air seems to be a good means of escape. It's not the only way, however. Life in communities or colonies, for many marine birds, is an additional protection factor, as is underground living for burrowing mammals (like the mole rat, which can live up to thirteen years, though it weighs only fifty grams).

The slowing of aging is more difficult to establish, always for the same reason, namely that mortality must be measured by age range. Nonetheless, whether it's a question of airborne or gregarious species, the preceding examples do suggest that less exposure to external dangers, resulting in lower mortality, is accompanied by slower aging. This conclusion is much less trivial than our former assertion. Delaying death does not usu-ally suffice to slow aging. Recall the disappointment of the Greek goddess of the dawn, Eos, who had been granted by Zeus immortality for her lover, the Trojan Tithonos, but forgot to also ask for eternal youth. Beset with the infirmities of extreme old age, Tithonos ended up transformed into a grasshopper. The myth actually reflects the current perception of aging as a phenomenon that appears even in the absence of any external aggression. Aging is caused by processes intrinsic to the organism, and not by external aggressions. Therefore, on an individual life's timescale, a less aggressive environment does not entail slower aging.

So what is to be made of the paradoxical link suggested between "ex-trinsic" mortality (which is inflicted by the environment) and aging? At least three experimental approaches support this link. The first two in-volve animals in the wild, with remarkably accurate results, given the complexity of such studies. Steven Austad conducted a study of a marsu-pial of the East Coast of the United States, the opossum (*Didelphis virgini-ana*), very similar in appearance to the hedgehog, but with highly prized gray fur and a long prehensile tail. Its life span generally does not exceed two years. Most females have only one reproductive year, after which they show many signs of aging, including atrophy of the sexual organs, re-duced fertility, weight loss, and cataracts. The young can be tagged dur-ing the two months they spend in their mother's pouch and their age mea-sured accurately by the length of their tails. Demographic studies are made even easier by their adult size: opossums are large enough to be fit-ted with long-term radio transmitter collars. In the wild, more than half the mortality of these rather placid animals is due to predation by birds and other mammals. Austad took advantage of an island population of opossums, which has lived separated from the mainland by several kilo-meters for a few thousand years.[17] This island is a peaceful haven for them, since it is home to none of their primary predators. This tranquil sit-uation has been reflected by striking behavioral changes. Unlike their brothers on the mainland, the island inhabitants do not rest in deep dens during the day, but instead often sleep out in the open.

Although the oldest females live hardly more than two and one-half years on the mainland, they live nearly four years on the island. The initial

[17] S. N. Austad, "Retarded Senescence in an Insular Population of Virginia Opossums (*Didelphis virginiana*)" *Journal of Zoology* 229 (1993): 695–708.

mortality rate there is lower, as one would expect. What is more surprising is that the island opossums also benefit from slower aging. There are twice as many "old" females (in their second reproductive year) on the island, and they are also fertile more often. In contrast, the average litter size has dropped from around seven and one-half to five and one-half, which reflects the link between longevity and reproduction seen earlier. Austad, after having studied other ecological factors that could influence the survival or health of the opossums, concluded that the slower aging of the island opossums could be attributed to their lower "extrinsic" mortality rate. This hypothesis goes counter to our everyday experience, which suggests that aging leads to death, and not the other way around. The explanation will have to wait until the next chapter, in which we examine this paradox in an evolutionary context. We will see that the mortal risks of life can shape aging—indirectly, but conclusively.

Although it is not specifically centered on aging, the approach of David Reznick and his collaborators is worth citing here.[18] It is exemplary for several reasons, and foremost for its experimental character. They worked on the guppy (*Poecilia reticulata*), a fish related to the carp and the goldfish. Rather than simply comparing existing situations, they made reciprocal transfers of populations between two rivers on the Antilles island of Trinidad that differ specifically in the type of guppy predator present. They were hoping to *prove* that there is a cause-and-effect relationship between environmental differences and certain characteristics of guppy populations—including age and size at maturity, fertility, and the number of descendants—that persists after several generations spent in a neutral environment, and must therefore be programmed into the genes. Due to their small size, guppies can be caught and transported in great numbers. However, their small size also complicates tagging, which is necessary to calculate mortality by age range. To solve this problem, the researchers injected dye into the guppies' tails so that the smallest individuals were downright flooded with a fluorescent product that binds to calcified tissues such as bones. This procedure required numerous controls, starting with looking at how these dyes affected the appetite of the predators, which could influence the mortality the researchers wanted to measure!

---

[18] "Life-History Evolution in Guppies (*Poecilia reticulata*): 6. Differential Mortality as a Mechanism for Natural Selection," *Evolution* 50 (1996): 1651–1660.

The scientific rigor required for this experiment makes the study fascinating to read. The time factor was the second inescapable practical difficulty for these researchers. It was not really feasible to wait for several thousand years to give the transplanted animals time to differentiate from their populations of origin, as was the case for Austad's opossums.

The initial experiment did last for eleven years, or about thirty generations. The results indicate that, in these guppies, hereditary characteristics of maturation and reproductive effort, which differ between the two rivers, are in large part determined by the type of predator they face. For the four major age ranges studied, and for both sexes, the mortality of the guppies in the river where the predators were more voracious was shown to be higher than in the other river. Though it has not been formally proven, it is tempting to consider this factor to be one of the main causes of the differences that appear between the two populations. What are the differences? Well, after a few dozen generations spent in the company of less voracious predators, the transplanted guppies reached their sexual maturity later, both in age and weight, than did their population of origin. The total weight of their first litter was also about roughly a quarter lower—a difference that disappears for subsequent litters. On the whole, we can say that the population that suffers from higher mortality also invests the most and the earliest in reproduction. These data do not provide any information on aging (this was not the purpose of the research), but they do provide another example of the flexibility of the major options regarding life and death, and also the crucial role of the mortality "suffered" by a species. Reznick's team has already conducted pilot studies suggesting that his system can be used to analyze aging, and that as aging evolves through the generations as a function of the environment, it may be possible to manipulate it in a virtually natural environment.

The third approach, developed by Stephen Stearns and his colleagues in Switzerland, consisted of imposing different extrinsic mortality rates on fruit flies in the laboratory.[19] In the "high mortality" population, the experimenters pitilessly eliminated nearly 90 percent of the flies twice a week, such that the probability of surviving more than one week was only 1 percent. To keep the density constant, they replaced the eliminated flies

[19] S. C. Stearns et al., "Experimental Evolution of Aging, Growth, and Reproduction in Fruitflies," *Proc. Nat. Acad. Sci. USA* 97 (2000): 3309–3313.

with their young descendants, raised in separate flasks, from eggs laid two weeks earlier by a sample from the corresponding population. Another population, produced from the same initial stock, was treated in the same manner, with one crucial difference: the probability of weekly survival was roughly 70 percent, since only 30 percent of the flies were eliminated each week. Intrinsic survival was tested in each generation on population samples in which flies were left alone to age quietly. Intrinsic survival improved in the population that evolved in an environment with lower mortality. However, its reproductive abilities at the youngest ages worsened. Thus, under laboratory conditions, which are both more artificial and more reproducible, these researchers obtained results similar to Austad's findings comparing two populations of opossums that had evolved over many generations in either hostile or favorable environments.

On the Frontiers of Immortality

The idea that death is inescapable and that it belongs to the very essence of life seems to be definitively laid to rest by the existence of a few species for whom natural death does not appear. Before presenting these species, a warning is in order. Any claim to immortality comes up against the necessarily limited duration of human observation. At best, we should talk about negligible aging—that which is imperceptible using our timescale.

We have already seen this phenomenon in this chapter, in the social insect queens. These queens also show, cruelly, that the absence of aging does not guarantee eternal survival. Death comes at the hand of the workers, their own daughters. The myth of Tithonos highlights, and rightly so, that only divine power can grant eternal life to aging organisms. But divine protection is just as necessary to eternally prevent death in organisms that do not age. Even though their potential life span is infinite, their probability of survival nonetheless diminishes with age, because of accidents, illnesses, and other hazards of life. In practice, this means that the records of successive longevity within a species that does not age are harder and harder to beat. They are a matter of pure statistics. The maximum recorded longevity depends on the mortality rate, but the more individuals who enter the race, the greater the chances that at least one of them will reach a given age.[20] . We should also note that an increase in mortality rate with age may be masked if the mortality rate is already relatively high at sexual maturity. This may incorrectly suggest the absence of aging. To avoid this type of error, it seems reasonable to exclude from the "imperceptible aging" category those species whose initial mortality rate is too high. In any case, their members would hardly have time to take advantage of their eternal youth.

Species with imperceptible aging exist in almost all the major subdivisions of the living world. Without contest, most of them are plant species. Conifers take top billing, with several species capable of living more than a thousand years: the sequoia, of course, but also the yew, the cedar of Lebanon, the ginkgo biloba, certain firs, etc. The certified record is close to five thousand years. And this is undoubtedly an underestimate, since usually some of the annual age rings are missing. Though nonconifers are rare among the thousand-year club, there are many that routinely live several centuries. It seems, in general, that the mortality of these multicentury-old trees does not increase, but the variation in their mortality rate with age is difficult to establish. Knowing the age of dead trees, which can be accurately evaluated, is not enough. It indicates only the number of deaths per age range, which should then be compared to the total number of living trees in that same age range to yield age-specific mortality rates. The only way to know the exact age of a tree, however, is to count the age circles in its trunk, which means cutting it down. The best indication of the absence of aging is that the fertility of these trees increases with their size. It is the "oldest" trees that provide the most seeds to form the next year's generation. When they die, it is due to external, mechanical causes, such as violent wind, fire, etc. This is in keeping with a development strategy that sacrifices the maintenance of virtually their entire bodies. As Leonard Hayflick noted, their bodies are composed mostly of dead cells; only a thin layer of the trunk, just below the bark, is living. The trunk can basically only resist or break. Looked at from a cellular standpoint, these large trees cut a sorry figure compared to birds or mammals with obvious but slow aging. Some of their nerve, muscle, and heart

---

[20] We have seen that the oldest individual of an eternally young humanity, that is, with constant annual mortality of roughly 0.1%, could reasonably expect to blow out his or her fifteen thousandth candle. But, on average, the oldest individual of a small city of six thousand inhabitants would have to be content with seventeen hundred candles.

cells, which appeared even before birth, can remain in good working order until the individual dies. Although the human life span is much shorter than that of an oak or a sequoia, the oldest cells of these trees are three or four times younger than any neuron of a human centenarian.

For a plant species to escape aging, humble grasses have demonstrated that the answer is not necessarily to build an imposing trunk. Some grasses show species-specific, constant mortality rates on the order of 1 percent per year. This rate theoretically allows one plant out of a thousand, on average, to live more than five hundred years. But there is even better, with some plants having life spans in excess of ten thousand years—yes, you read correctly!—leaving even the conifers in the dust. To tell the truth, these absolute records smack of cheating, since they are based on asexual propagation, by simple "budding" followed (or not) by fragmentation. The budding may take several forms. The aspen, for example, a tree related to the poplar, propagates by sending off root sprouts. There is an aspen in the Rocky Mountains that now covers nearly twenty acres. Other plants propagate by rhizomes, underground stems from which other individual stems grow up to the open air. The size of the resulting bushes and how fast they propagate can be used to establish their ages. The record seems to be held by a wild blueberry, dating back more than thirteen thousand years (and having a diameter of nearly two kilometers), far older than any fern. These bushes are clones, that is, populations composed of genetically identical individuals.

Farmers have long taken advantage of the ability of plants to reconstitute an entire organism from a small fragment, by graft or cuttings. More recently, biologists have reduced this fragment to its simplest expression. They now know how to regrow an entire plant from just a single cell taken from an adult plant.[21] It was more than thirty years ago that tobacco and carrot plants had their Dolly, the sheep that raised such a controversy in 1997 as the first successful example of this type of cloning in an animal. If cloning an animal is more difficult, it is doubtless because the separation between somatic cells and germ cells is more marked than in the plant world. Year after year, iteroparous plants have to regenerate their sexual organs—the flowers—in their zones of growth. Regardless of the age of the plant, these zones retain the ability to produce sex cells (pollen,

[21] This of course involves cells that do not normally have this function, unlike sex cells.

ovules) and therefore do not belong to a specialized, separate line, as in most animals. The germ-soma dichotomy, introduced by Weismann, seems much less clear for plants. This particularity is not without consequence for the theories on aging discussed in subsequent chapters.

Reproduction by budding or fission, although it is called "vegetative propagation," is also encountered in relatively simple multicellular animals, such as flatworms, annelids, ascidians, sea anemones, corals, and hydras. Most of these creatures do not clearly manifest clonal aging, which is to say that the population engendered by a single organism does not seem to have any absolute limit in terms of time. But the potential immortality of the clone does not require the individuals that compose it to be immortal. In fact, individual aging (specifically, increased mortality with age) has been established, at least in certain cases, though with less certainty than for baker's yeast.

This humble microorganism can provide valuable food for thought. Capable of sexual reproduction, it ages even when it reproduces asexually. The daughter yeast is a bud, which remains smaller than the mother for quite a while. Mother and daughter can be identified by differences in size, then separated to monitor the successive reproductive cycles of a given mother. The fundamental observation, made in the late 1950s,[22] is that this number of cycles is limited. Certain "individuals"—made up of a single cell, remember—stop before twenty cycles, and no individual exceeds fifty cycles. After that they are definitively out of the race, and most of them lyse. The status of the few "survivors" is uncertain, which once again shows how much trouble biology has in defining "natural" death, other than by default and its irreversibility.[23] From the standpoint of their contribution to future generations, these yeast mothers that no longer bud are certainly dead. In contrast, the "newborns" start over at zero, unless their rank in the "brotherhood" is very high. Thus, although each yeast serves its time and then dies, the population is immortal and shows no signs of clonal aging.

These results are surprising. But there's even better. If we express the age of a yeast as the number of reproductive cycles completed, then the

[22] R. Mortimer and J. Johnston, *Nature* 183 (1959): 1751.
[23] A similar ambiguity surrounded the bacterial spores we brought up early in chapter 2. Those that resumed dividing were, if not alive, at least viable. By definition, dead cells are those that do not "resuscitate."

probability that it will die (stop budding) increases exponentially with its age. The doubling time (less than two days) is among the shortest of all. The actuarial calculations of Gompertz also apply to a microorganism! And not just any microorganism, from the point of view of molecular biologists, since yeast is one of their favorite organisms, along with the bacterium *Escherichia coli*, the fruit fly, the mouse, and a few others. It offers gerontologists even more benefits. First, the "chronological" age of yeast can be read by counting the number of scars left by the buds. Second, the time between two consecutive buddings is an excellent and unique biomarker of individual aging. This time interval increases exponentially with the number of generations and can be used to accurately predict a given individual's death several generations in advance. By combining classical genetics and molecular biology, researchers have been able to identify the genes that help determine the longevity of yeast.[24] It's a good bet that yeast, whose genome (one of the smallest of all eukaryotes) was completely deciphered in 1996, will contribute at least as much to research on aging as it has already contributed to our understanding of cancer.

These natural deaths, unicellular and asexual, show that aging cannot be simply the price to pay for multicellular differentiation, as the anatomist Charles Minot asserted, or the "inevitable toll of sexuality."[25] Doubtless to console ourselves for this supposed toll, another preconceived notion proposes a more positive link between sex and death—that of "resurrection via sex."[26] Here again, yeast proves this assertion to be false. A yeast produced sexually, by the fusion of two yeasts of opposite sexes, is, in fact, as old as the oldest of its parents. Thus, sex does not necessarily reset the counters to zero. Biologists, in their attempts to prove that sex has a rejuvenating effect, have often looked for examples among the ciliated protozoa such as paramecia. Graham Bell described these attempts and the controversies they gave rise to in his book *Sex and Death in Protozoa*, with the eloquent subtitle: "The History of an Obsession."[27] The word *obsession* is not too strong, given experiments that lasted up to twenty-two years, in which researchers monitored the successive divisions of an initial cell daily under a microscope, singling out one cell by

---

[24] These experiments will be described in chapter 8.
[25] J. Ruffié, op. cit., p. 270.
[26] J. Ruffié, op. cit., p. 260.
[27] Graham Bell, *Sex and Death in Protozoa: The History of an Obsession* (Cambridge: Cambridge University Press, 1988).

separating it from all the other cells produced by division, renewing the culture medium several times per week, and so forth. The first observations showed that these beings, despite being unicellular, ended up degenerating en masse after a certain number of divisions, with alterations that could resemble alterations in the cells of an old multicellular organism.

This clonal aging was prevented, or even reversed, in cultures where sexual pairing (the technical term is conjugation) of cells could take place. Is this not an equivalent of aging in multicellular organisms, and at the same time proof of the role of sexuality, namely to erase the traces of aging with each generation? Clearly, the stakes were worth the efforts invested. Unfortunately, the results turned out to be disappointing. First, because of the diversity of the situations, since, as in plants, certain species or lines of protozoa do not experience clonal aging, even in the absence of conjugations. And second, because no truly significant similarity could be established between clonal aging and aging in whole organisms, or even in cell cultures created from whole organisms.[28]

The studies summarized by Bell were not in vain, however. They provided important clues as to the role of sexuality. Basically, sexual reproduction seems to prevent the accumulation of harmful mutations in the genome of the species, over the generations. The formation and subsequent fusion of sex cells is accompanied by countless exchanges and reshuffling of chromosomes, in what is often referred to as the genetic lottery. Winning tickets—genomes purged of at least some of the harmful mutations—reappear at each generation. Their winnings are reflected by more numerous offspring, which help reduce, inversely, the proportion of individuals that carry harmful mutations. Thanks to sex, the genome of a species does not age. Perhaps this may assuage our initial "obsession" a bit, if we consider that sex results in a sort of collective rejuvenation.[29]

Let's return now to a world that is a little closer to our own, and look at multicellular animals that reproduce sexually. We have already mentioned in passing examples of potential immortality in invertebrates, such as insects (social queens, perhaps certain coleoptera) and parasitic worms. And there are many others, including bivalve mollusks that are considerate enough to accurately indicate their age on their shells by means of an-

---

[28] See chapter 5.

[29] It remains to be explained why some species seem to be able to do without sex, but that is another story, one about the origins and perennial nature of sexual reproduction (see the article of P.-H. Gouyon et al., *La Recherche*, January 1993).

nual growth rings. Many species live for more than a hundred years. As always, the change in mortality as a function of age is less well established, but the available data suggest that aging is negligible. In clams, the production of sex cells increases regularly with the size and age of the organism. We will see that this is an indirect argument in favor of potential immortality.

The same pattern holds true for crustaceans such as barnacles and, in an even more spectacular way, lobsters. In female lobsters, egg content increases with age almost faster than body size! Record catches exceed twenty kilograms, and no absolute limit has yet been discovered. Growth is accompanied by shedding, initially twice yearly, which fully renews the shell. This process eliminates problems caused by wear, but also prevents precise age measurements. But based on known growth rates, the largest lobsters ever caught could easily have been more than one hundred years old. As for sea urchins, animals a little more closely related to us (despite appearances), their mortality rate actually seems to decrease as they grow larger, according to some studies, perhaps because their spines get bigger and more of a deterrent.

The preceding examples involve only relatively "primitive" species. Vertebrates that show negligible aging would provide better food for thought on human aging, and our dreams of immortality. Fish are in the best position to make a claim to immortality, especially, as noted earlier, since the age of a fish can be accurately established. Several species have an enviable life span, even longer than the legendary carp.[30] The record sturgeon, at 154 years, is followed closely by scorpion fish, which live from ninety to 140 years depending on the species. Flat fish such as flounder, sole, and halibut commonly live more than fifty years, and perch live relatively long as well. As always, mortality rates are difficult to measure. In general, for these species, no obvious increase in mortality with age has been observed except in perch and sturgeon (clearly, exceptional longevity is not synonymous with eternal youth), and even here the increase is

[30] Even in scientific circles. Very naturally, it was the carp that was chosen by the very serious journal *Nature* in 1982 for an April Fools' Day article entitled "The Elixir of Life" (supposedly written by the clever pen of Alex Comfort). Supported by experimental details and bibliographic references, the article described the purification of an imaginary protein from carp intestines, longevin, so called because of its effects on the longevity of laboratory mice. Unfortunately, longevin also had a few unwanted side effects, such as causing the fur to be replaced by scales.

less rapid than in humans. Indirect data sometimes support the hypothesis of an absence of aging—as in lobsters, which exhibit apparently unlimited, though progressively slower, growth; female fertility that increases with age; and an absence of obvious pathologies on dissection of old specimens. But this hypothesis is not unanimously accepted.

Amphibians rank rather high in the longevity contest, with a fifty-five year life span for the Japanese giant salamander. The clawed toad lives at least fifteen years. The few data available do not indicate any marked aging for this animal, a favorite of research labs (though not so much in labs doing research on aging, which is perhaps a shame). In general, the reproductive functions of salamanders, frogs, and toads do not decrease, and in some cases they can even increase, up to the oldest ages studied.

Reptiles also present candidates for potential immortality. Several snakes, including the asp viper, can live for more than twenty years without any notable increase in mortality. Some continue to grow, although more slowly, throughout their entire life. The number of eggs laid also tends to increase with age. Turtles are widely considered to be a symbol not only of slowness, but also of longevity. Is their claim to longevity entirely deserved? Although many species live longer than twenty-five years and some giant turtles have indeed lived to age seventy in captivity, the record of 150 years claimed for a turtle in Mauritius is less certain.

The situation is less encouraging in the two classes of homeothermic vertebrates, birds and mammals—not that their "warm blood" is a direct cause of death. Birds, whose internal temperature is on average several degrees higher than that of mammals, and whose metabolism is faster, live longer than mammals (at a given weight). For that matter, they present serious candidates for negligible aging. The mortality rate of condors and petrels, for example, does not seem to increase for at least several decades after sexual maturity. The condor's life span exceeds seventy-five years. The reproductive success of birds often increases with age, doubtless thanks to their increased experience in building nests and raising their young. In contrast, there is not a single mammal that does not age. This absence of exception is not at all surprising, given the relative similarities of the aging across this class of vertebrates. Perhaps their organization necessarily implies natural death. But even if this pessimistic hypothesis proves to be true, it does not rule out the possibility of interventions that could slow down aging. After all, human aging is already among the slowest of all mammals, so why shouldn't we be able to slow it down even further?

Death as an Option

This trip through the land of "natural" death, by showing us just how diverse natural death really is, has served to dispel some preconceived notions, or at least to highlight their limitations. Decidedly, there is no obligatory link between death and sexuality, or between death and multicellularity, for example. But the main conclusion is that death is not necessarily the ultimate fate of all beings. Although it is still the inevitable end to any individual life that is subject to the hazards of sickness, predation, or fatal accident, there is no greater law that inexorably dooms all living beings to age and die, even if each particular species strictly obeys its own rules in the matter. From mayflies to giant redwoods, it takes all kinds of choices to make a world. Still, these choices are not made randomly. Despite the incredible diversity, the end results are relatively similar. In all mammal species, the females give birth during an entire lifetime to almost the same average number of descendents (between nine and twenty-four, depending on the species), whereas among the same species life spans vary one-hundred-fold and weights vary four-million-fold!

The rate of aging and the longevity characteristic of a species seem to result from a subtle equilibrium, a series of compromises that simultaneously determine many other properties of this species, under the pressure of ecological constraints that are just now starting to be identified. The next chapter attempts to show that these compromises are not made so much for "the good of the species" but rather that they depend on the prospects of each individual organism. Without making conscious, deliberate choices, the organism follows a genuine existential strategy, programmed into the genome of its species and accessible to human knowledge. Each of these strategies defines what biologists call a life cycle or life history, specific to a given species. Aging is just one component of the life history, inextricably linked to and evolving together with the other components.

CHAPTER FOUR

WHAT'S THE POINT OF DYING?
Evolutionary Theories on Death

The concept of adaptation should be used only as a last resort. . . . It would be absurd to recognize an adaptation to achieve the mechanically inevitable.

G. C. Williams

The preceding chapter touched only briefly on the incredible diversity of strategies, life cycles, ways of aging, and death. Is it possible to construct a theory that accounts for this diversity?

Aging is in most cases an artifact of domestication, or civilization, much more than it is a product of nature. Indeed, in most species in the wild, individuals have very few chances of reaching the "normal" end of their existence through aging, since they are much more likely to be killed by a predator than to die a "natural" death. Aging seems to have very little opportunity to occur. And what use would it be anyway, since the result of aging is to increase the mortality rate with advancing years? A process that accelerates the elimination of individuals after a certain age obviously does not, in theory, seem to benefit any of those who are subjected to it. Does this mean, as the naturalists of the eighteenth century believed—along with quite a few authors even today—that the group, the species, or the entire living world as a whole has something to gain from the process?

The answers, as we saw briefly at the beginning of the book, may be found by looking at the mechanisms involved in the evolution of a species. The answers may not be definitive, but they solve a number of paradoxes and dispel quite a few preconceived notions. In particular, we will see that the sudden limitation of life span that results from aging can be explained without implying that death and aging represent a beneficial adaptation for either the individual or the species. This demonstration is fundamental, because it calls into question some very widely held notions about the role and "value" of death—the consequences of which reach far beyond the field of biology.

Prologue: An Overview of the Theory of the Evolution of Species

Before getting into the thick of the subject, we would like to sketch out the general background (evolutionary theory) which gives meaning to the extraordinary variety of the living world, and the striking ways in which species adapt to their environment. This conceptual framework began a little over two centuries ago, when the Biblical explanation of Divine Creation, a world fixed once and for all, no longer satisfied certain naturalists. Jean-Baptiste Lamarck was one of the first to propose replacing the word "fixism" by "transformism" (the word "evolution" did not appear until later). What could have been simpler to explain the transformation of species, and the appearance of new ones, than the hypothesis formulated by Lamarck at the very beginning of the nineteenth century, known today as "hereditary transmission of acquired characteristics"? One of his famous examples is the long neck of giraffes. According to Lamarck, giraffes developed long necks because of the sustained efforts of successive generations that were obliged to look for food at the tops of trees that were often scarce in the African savannah. Each giraffe, in attempting to stretch its neck to reach food, supposedly modified its morphology slightly, and this modification was transmitted to its offspring according to the principle of hereditary transmission of acquired characteristics. Given their environment, giraffes obviously had everything to gain. Their necks grew longer from generation to generation until a suitable length was reached.

During the 1880s, August Weismann, one of the founding fathers of genetics, after considering the fact that the longevity specific to each species must be programmed into its heredity, contradicted Lamarck's theory (which was very widely accepted at that time). We can see, intuitively, where Weismann ran into problems when he tried to understand the evolution of longevity in terms of the heredity of acquired characteristics. An individual can die only once. When he or she has acquired this "characteristic" (dying at a certain age) he or she is no longer in a position to transmit it to his descendants.[1]

Recall that Weismann postulated a radical separation between what he called the germ (the sex cells, which carry the characteristics transmitted to descendants), and the soma (the rest of the organism). The soma expresses the characteristics received from the preceding generation via parental germ lines. In more modern terms, parents transmit genes to their children via their sex cells, and the genes thus transmitted play an essential role[2] in the development of the child and its adult characteristics. According to Weismann, and most modern geneticists, this influence of the genes on the soma is a one-way street. What happens to a certain part of the parental soma—say the neck, in the case of the giraffe—cannot affect the genes that influence the characteristics of this part of the body—the length of the neck. The genes that a baby giraffe inherits are not changed by its parents' efforts, or the lack thereof. Weismann based his conclusion above all on the example of longevity. "I cannot imagine," he wrote, "any way in which the somatic cells could communicate this assumed death to the reproductive cells in such a manner that the somatic cells of the resulting offspring would spontaneously die . . . in the same manner and at a corresponding time as those of the parent."

Weismann, basing his findings essentially on these types of theoretical considerations, stated this general principle: the characteristics acquired during the lifetime of an individual cannot be transmitted to its descendants. The hereditary nature of acquired characteristics was finally ruled out. This "Weismannian revolution" provided a fertile conceptual framework for the scientific study of heredity, even if it obviously had its limits. The ideas of Weismann also contributed greatly to the theory of evolution generally accepted today, based on the works of the English naturalist Charles Darwin. This theory is very different from Lamarck's. Let's go

[1] Weismann raises a similar objection about "the many instincts that are exercised only once in a lifetime" such as the mating dance of the bee or the building of a cocoon. The individual does not have the chance to improve on his instincts before transmitting them to his descendants. Charles Lenay drew our attention to the fact that Weismann's thinking was greatly influenced by his reflections on aging (*La Découverte des lois de l'hérédité* [Paris: Presses Pocket, 1990], pp. 197–199).

[2] This does not mean "exclusive." See M. Morange, *La part des gènes* (Paris: Odile Jacob, 1998).

back for a moment to our famous giraffes. At each generation, the primitive giraffes with longer necks had an advantage, because they could nourish themselves better than the others. Up to this point, Lamarck would agree. But it is the genes that the giraffes received from their parents that determine the length of their necks, not their efforts. These genes themselves are in no way affected by the height of the trees or the quantity of foliage available, or by the use the giraffes make of their necks.

If their necks are not all the same length, this is because of the genetic variability that exists in all species in the wild. During evolution, certain ancestors of present-day giraffes inherited genes that were a little different, which gave them necks that were a little longer than those of their fellow giraffes. These individuals that could obtain food more easily had more offspring, on average. We say that they were selected, hence the term natural selection. Their genes—including those that conferred a longer neck—were present in greater quantities in the following generation. On average, the individuals of this next generation benefited from having a longer neck than giraffes in the previous generation. And so forth until they achieved a neck that was, perhaps not optimal, but at least remarkably well adapted to a food source that few other herbivorous mammals could reach.

This concept of evolution, often called neo-Darwinism, is currently the most widely accepted theory.[3] According to this theory, natural selection can be represented as the driving force behind evolution, a force that is fueled by genetic variability. Giraffes' efforts are not aimed at lengthening their necks, but at producing offspring[4]—that is to say, effectively transmitting their genes, including those that determine neck length. Unlike the poor mortal organisms that we are, genes are in a manner of speaking immortal, since they are passed on from generation to generation.[5]

[3] Darwinism because it comes from the works of Darwin, who was the first to propose the idea of natural selection, and *neo*-Darwinism because it does away with the notion of inherited acquired traits, which Darwin accepted. According to the historian A. Pichot, this theory should be called Weismannian. In all strictness, today we should speak of the synthetic theory of evolution, which is neo-Darwinism supported by genetics (since genes had not been discovered in Darwin's time, or when Weismann published his first works) and by population biology.

[4] Quite a job in itself!

[5] We mentioned this in the first chapter as the "immortality of the germ line" with the exception of mutations and other alterations in genetic material. Such mutations and alterations are rare, occurring in roughly one per million nucleotides per generation, because the DNA replication and repair systems were themselves selected for their precision. Even

Aging: An Inescapable Side Effect of Natural Selection

Now here is the next question: how could natural selection have created organisms that age until death occurs? A living organism is obviously not just a machine that wears out over time. It is constantly being renewed, unlike a machine whose parts are built once and for all (individual parts can be changed, but there comes a day when the whole machine has to be replaced). In fact, from a molecular standpoint, we are a completely different person every few months. Of course, the ceaseless self-renewal and self-repair mechanisms of the living may sustain irreversible damage. But even if such damage does occur, this does not explain aging. Natural selection could favor those mechanisms that resist damage, thus increasing life spans.

Allow us to reiterate the fact that death—and prior to death, aging, which increases the probability of dying—can hardly be considered advantageous to the individual. However, these limits to individual existence are in various ways written into the genome of practically every living species, even though individuals rarely reach the maximum life span of their species in the wild. How could natural selection favor a programmed elimination process that hardly ever comes into play? Is it not astonishing, as George Williams wrote in 1957, that after a multicellular organism successfully performs "a seemingly miraculous feat" of development, it "should be unable to perform the much simpler task of merely maintaining what is already formed"?

Of course, it might be concluded that the elimination of old individuals benefits the species. But as we pointed out in chapter 1, this argument is extremely precarious. Despite its weaknesses, this theory has always been considered to be very attractive, and still comes up today in many works for the general public and in a few scientific articles. Weismann himself used this argument. In 1881, however, he presented another view. He believed that because the immortality of the individual was not necessary for gene transmission, or for the preservation of the line, there was no reason for immortality to be preserved. Indeed, once its hereditary character-

---

on a geological time scale, certain genes have been preserved in such a way that we are able to trace the genealogy of life almost to its origin, 3.5 billion years ago. But let's insist that without mutations evolution would not be possible.

istics are transmitted, the individual inevitably ends up as an "accessory appendage." What would be the point in keeping it alive? Generally speaking, if a function—in this case, immortality—is not useful, it is lost: "Any function and any organ disappears as soon as it becomes superfluous for the preservation of the species."[6]

This statement lies at the heart of Weismann's 1891 theory on regressive evolution, which is perfectly illustrated by cave-dwelling animals. Their vision is weak or they may even be sightless, in which case they have only vestigial eyes. In these and other species that do not go out into the daylight, sight organs have gradually regressed. This is not a Lamarckian effect due to lack of use, rather the function is lost because natural selection, which normally plays a stabilizing role, has stopped acting on this function. Natural selection does not drive evolution only in the "favorable" sense (to avoid using the controversial term progress). It also preserves previously accumulated advantages. Genetic variations that weaken these advantages appear in every generation, and in every generation they are eliminated by natural selection. But when a certain function becomes useless, it no longer matters whether it is effective or not. The selective value of this function (vision for cave-dwelling animals or immortality of individuals for virtually all species) becomes zero. From generation to generation, the genetic defects that reduce the effectiveness of the function can spread freely in the species. As Weismann said: "Death became possible, among multicellular organisms in which the somatic and reproductive cells are distinct, and we see that it has indeed taken hold."[7]

Let's go back to the silkworm moth studied by Metchnikoff. This organism illustrates Weismann's reasoning perfectly. Unlike most moths, but like many species of aphagic insects,[8] the silkworm moth is incapable of feeding in the adult state. It has no mouthparts enabling it to ingest food, either liquid or solid. The adult lives, for the two to three weeks of its existence, on the fat reserves built up by the caterpillar. Since these reserves are enough to fuel reproduction, why should the moth need to feed, even though it is doomed to die once the stores run out?

---

[6] Quoted by Kirkwood and Cremer (op. cit.). Present-day evolutionists would replace the phrase "the preservation of the species" by something like "the reproductive success of the individual."

[7] Weismann, *La Découverte des lois de l'hérédité*, op. cit.

[8] See previous chapter.

This relatively rapid end is a result of internal constraints programmed into the organization of the individual, which is why Metchnikoff saw this as a good model of unquestionably natural death. However, things are not as simple as they seem. First of all, we could "force-feed" the silkworms. They do have a complete digestive tract, and would be able to assimilate the nutrients. We might well say, then, that their death has an external cause—the absence of appropriate feeding assistance—which is almost accidental rather than natural. Of course, this argument is extremely far-fetched, since force-feeding the moths is not natural at all. It does, however, show that determining the basic mechanisms that limit the longevity of a given species can be quite tricky. For that matter, Metchnikoff said that "this death cannot be attributed to the desiccation of the tissues. . . . Nor does the silkworm moth die of hunger . . . [since] the considerable fat reserves of our moths are still partly unused at the time of death." He also noted the production of toxins that build up in the urine, which is held in the bladder for almost the entire adult life.[9] Now things get really complicated, since the moths seems to be overly determined to die a natural death. In fact, the adult form of this species doesn't need urine detoxification mechanisms for exactly the same reason it doesn't need food. Once the continuity of the line is ensured by the formation of the next generation, the fate of the individual is unimportant. Adult moths have absolutely no need to feed or to eliminate their waste.

We seem to have gone from a utilitarian to a somewhat nonutilitarian concept of death. It's not that natural death is useful, but rather that there is no reason to prevent death. So, could the loss of the immortality we yearn for so deeply merely be due to the fact that it is useless? Not only that, and not always, according to Weismann. He believed that the gradual loss of an organ during the evolution of a species is all the more likely to happen, "inasmuch as other organs that are of importance for the life of the species will gain what the functionless organ loses in size and nutrition." Etienne Geoffroy Saint-Hilaire (another important figure in our Museum of Natural History) proposed a similar idea, under the name of "law of balancing of organs." Just as blindness is not in itself useful to a cave-

[9] Metchnikoff's research led him to recommend a yogurt-based diet to extend human life. He believed that the lactic fermentations would combat the toxins that accumulated in the intestine. This reasoning may seem ridiculous today, but do our current health magazines provide better advice?

dwelling animal, death is not directly useful to a living organism. But the loss of immortality, like the loss of sight, may free up resources and open unsuspected prospects. (We will look at some examples of this later.)

Thus, more than a century ago, Weismann suggested that the theory of evolution could account for natural death without implying that it provided any direct advantage. Death is not selected on its own merits ("making room for the young" or other arguments). The division of labor between the germ cells responsible for reproduction, on one hand, and the somatic cells that make up the rest of the individual, on the other hand, has freed the individual from the constraint of eternal life. This opened up a new field of possibilities, including the appearance of natural death. Once immortality has become useless, it is doomed—and doubly so, if, in addition, the loss of immortality benefits the species.

Today, Weismann's ideas can be expressed in more scientific terms, using concepts of molecular genetics and population genetics. As they say in math circles, let's reason "by contradiction."[10] Let's assume there is a species in which the individuals are not subject to aging. Let's also assume that their fertility does not decrease with age, but neither does it increase (this condition is essential for the rest of the reasoning). These individuals have, in theory, a maximum life span that is infinite, and they remain eternally young. Does this mean that they are eternal? No, for we can reasonably suppose that some day or other they will have an accident, or they will encounter a predator or a virus that will bring their existence to an end; the older an individual is, the luckier it is to still be alive. Although reproductive performance remains constant, the probability of procreating decreases with age, simply because the probability of having died increases. In other words, the oldest individuals contribute less to the future population via their offspring because their age class is less numerous. Even though we are dealing with individuals that are potentially immortal, who do not age, and whose fertility remains constant, the procreative weight of the "old" individuals decreases with their numbers.[11]

This point having been established, let's continue our reasoning. In all populations, there is genetic variability, as most genes have several pos-

---

[10] Our statements here are based essentially on the theories of P. B. Medawar and G. C. Williams. See P. B. Medawar, *An Unsolved Problem of Biology* (London: K. K. Lewis, 1952), and G. C. Williams, "Pleiotropy, Natural Selection, and the Evolution of Senescence," *Evolution* 11 (1957): 398–411.

[11] In the absence of aging, the decrease in number of individuals is exponential, as in the case of radioactive atoms . . . or dishes (see chapter 2).

sible forms (or alleles). Mutant forms appear all the time, spontaneously.[12] Throughout this chapter, we are referring only to mutations that occur in the germ line cells, and which can therefore be transmitted to descendents.[13] Usually, natural selection eliminates mutations that reduce the number of descendants or their ability to survive and reproduce. A balance is established between the pressure of harmful mutations (in other words, how frequently they appear) and the pressure of selection.[14] Consider a harmful mutation that reduces the probability of survival to sexual maturity by 50 percent. Overall, the individuals that inherit this mutation will have two times fewer descendants than the others. Now take a mutation that reduces survival by 50 percent but only beyond an advanced age, which is reached by only 10 percent of the population of the species in question. Nine out of ten carriers of this mutation will be dead before it manifests itself. As for the tenth individual who remains, he or she will already have had numerous offspring. Only a small fraction of his or her total descendants will be reduced by half.

Thus, a given mutation will be more harmful the earlier its effects are manifested, and natural selection will tend to effectively eliminate mutations that cause defects early in life. In contrast, it may easily allow a mutation whose negative effects appear later in life, when an individual has already had most of its descendants. A mutation that has a harmful effect on reproductive abilities or the survival of only the oldest individuals will therefore be able to spread in the population, through the generations. This mutation marks the appearance of aging in our hypothetical species, which initially did not age. What's worse, such a mutation will snowball, by further decreasing the procreative weight of the oldest individuals. Natural selection will be even less effective at eliminating any new mutations of this type. Under these conditions, a species whose members do not age would not keep this privilege for long.

This reasoning can account for one general observation: In many

---

[12] As a very rare but inevitable consequence of the DNA replication processes, and of various aggressions, specifically ultraviolet rays, X-rays, and chemicals.

[13] We will see in the next chapter that DNA damage in *somatic* cells doubtless plays a role in aging. It is important to distinguish between these two levels of explanation. Only *germinal* mutations are involved in the evolution of species.

[14] A simple mathematical formula describes this balance: $p = u/s$ where p is the proportion of carriers of a given harmful mutation, u is the rate of de novo occurrence of the mutation, and s is a selection coefficient between 0 and 1 that is a measure of the corresponding "disability." If s is close to 0 (very weakly harmful effect), the formula does not work, because the mutation can spread almost freely, until it invades the entire population.

species, the age at which individuals die the least—for humans, approximately fifteen years of age—corresponds roughly with the age of sexual maturity, the age at which "reproductive value" is the highest.[15] Many dangers lie in wait for newborns and young immature individuals. As they avoid the risks and advance in age, the probability that they will some day procreate increases with approaching sexual maturity. Natural selection is particularly effective for anything that affects an organism that has just reached maturity, since the reproductive potential of a future adult hangs in the balance. Both before and after that age, natural selection is less effective. This may be why the mortality rate decreases gradually with age until sexual maturity, and then increases. Again, this does not mean that procreation in itself always induces aging, or that those who abstain from sex do not age. Except for the cases of sudden death described in the previous chapter, the link between reproduction and aging is much more indirect. Put simply, reproductive maturity marks the moment when natural selection begins to be less vigilant.

In conclusion, our initial situation consisting of a species of individuals that do not age, experience no natural death, and have a theoretically infinite life span is completely unstable from an evolutionary standpoint.[16] But this does not mean that evolution "programmed" aging or that aging is directly selected because it is an advantage in itself. As a general rule, aging is much more an inescapable side effect of evolutionary processes, of the way natural selection works—namely, it is less severe on deficiencies that appear later in life than on those that affect young individuals.[17] Natural selection cannot preserve the potential immortality—the absence of aging—of a species that is not completely protected from the mortal hazards of life.

Of course, these hazards are not the same, either in quantity or quality, for all species, depending on their lifestyle, their environment, and so forth. A fly's chances of survival, even if it remains eternally young, are

[15] The British mathematician R. A. Fisher, one of the founders of modern genetic theory, introduced this notion in 1929, when he said that the force of natural selection should be proportional to reproductive value.
[16] Remember that this conclusion is based on only two relatively non-restrictive conditions: that the initial organisms are likely to die accidentally and that their fecundity does not increase with age.
[17] From an evolutionary standpoint, this theory accounts for the appearance not only of aging as such (see also following paragraph), but also age-related diseases that affect a large part of the population. Thus in humans, Alzheimer's disease, a particularly disabling neurodegenerative disease, affects one out of every four people over the age of eighty. This proportion is never encountered for diseases that affect young people.

much lower than an elephant's, even an aging one. Natural selection then succeeds in slowing or delaying aging, over the course of evolution, more effectively for elephants than for flies. The less time we are likely to survive, due to predators or other external causes or mortality, the more quickly we age.

This result is just one of the fascinating predictions of the theory largely credited to Medawar and Williams. Like many others, this prediction is roughly in line with what we know about aging. The aging of flies, measured by an increase in the mortality rate with age, is effectively faster than that of elephants. One of the most convincing studies in this area is Austad's work on the opossums of Virginia. After a few thousand generations in the protected environment of an island with no predators, these charming marsupials age two times slower than their cousins on the mainland. Another example is the exceptionally slow aging of bats and most birds[18] compared with related species that cannot protect themselves by flying away.

Evolutionary theory also predicts that aging must not systematically result in the "failure" of any one organ long before all the others. Why? Because natural selection would work harder at eliminating this form of premature aging, since reproductive value is higher at that age. This prediction seems to be borne out: in cases where wear is measurable, the various organs statistically wear out at more or less the same rate.[19] This does not preclude a wide individual variability within each species: in one the first to go is the heart, in another it is the liver, in a third it is the brain, and so on.

Williams imagined the case of species whose fertility continued to increase throughout life. Such species would not meet the condition of constant fertility required for our proof by contradiction of the inevitability of aging. Indeed, the increase in fertility might compensate for the gradual decrease of the number of survivors with age, which was the other condition of our reasoning. In this case, natural selection would be just as effective against mutations appearing late in life as it is against those appearing early, so that aging should be negligible, or even non-existent. Here again, the prediction is in line with the facts presented in the preceding chapter. In several species of fish, in which the females produce more and

[18] See previous chapter.
[19] It is difficult to know with certainty, if aging occurs too quickly (for insects, see chapter 3).

more offspring with age, it is difficult if not impossible to find obvious signs of aging. Likewise, the maximum life span of many trees is open-ended. In general, it takes a major disaster, like a forest fire, an earthquake, or lightning, to make a sequoia stop growing and die.[20]

This link between the ability to grow and longevity was also observed by Weismann and Bidder. After realizing that animals that grow indefinitely do not show any signs of senescence, they suggested that aging was a side effect of mechanisms whose primary purpose was to limit the size of adults.[21] Williams's interpretation is completely different. He noted that the indefinite growth of adults, whether in the case of trees or fish, is related to the continued increase in their fertility, which allows them to increase their reproductive value with age. In this case, natural selection does not "lose interest" in old individuals. During the evolution of such a species, aging processes do not appear or are imperceptible. These two points of view are very different. If we follow the ideas of Weismann and Bidder, eliminating whatever is limiting an individual's adult size would immediately increase its life span. But according to Williams, there is generally no direct link between the mechanisms that curb growth and those that reduce longevity. There is simply a relation of timing. The limitation of life span by aging appeared during the evolution of species with limited growth because their fertility does not increase with age. Artificially prolonging the growth of an individual beyond its normal limits will not make it live any longer. We will see that the scales are tipping more in favor of Williams's interpretation.

At the opposite end of the spectrum from species with indefinite life spans are animals (spectacularly exemplified by some salmon species), as well as many plants, whose individuals conclude their development with a single reproductive cycle and die immediately thereafter. Before Williams, what was the standard explanation? If they reproduced only once, it was because they died afterward, having accomplished "their duty towards the species." The evolutionary viewpoint completely overthrew this reasoning and made the notion of duty to the species superfluous. It is because members of these species had only a very small chance of reproducing a second time—because of predation or harsh seasons, for

---

[20] However, to qualify this well-known statement on the longevity of trees, let us add that most of what they are (most of the trunk and branches) has been dead for a long time when the fatal accident occurs.

[21] See chapter 1.

example—that natural selection favored the individuals that would be most effective during their first attempt, without worrying about what might happen to them afterward. Having seized their only chance and "given it their all," they died of exhaustion immediately afterward. This abnegation, controlled by their own hormones, is not a sacrifice on the altar of the species but rather one of many possible ways of optimizing their own individual reproductive success. Given the constraints specific to their species, these individuals gain an advantage by betting everything on a single reproductive cycle, because they stand to lose a great deal by scrimping in order to reproduce a second time. This considered risk is what has oriented the evolution of their life cycle, to the point that this second time has become physiologically impossible, although at the outset it was only improbable.

In fact, in Pacific salmon, we encounter the full range of possibilities, from species whose individuals inevitably die after they complete the long swim upstream to reproduce, to the species—commonly called trout—whose members occasionally survive after reproducing. For freshwater trout, survival after reproduction is even common. The accelerated damage that accompanies reproduction is irreversible and fatal for the former species, and more or less reversible for the latter. In the plant kingdom, *Lobelia telekii* and *Lobelia keniensis* illustrate this option. Everything seems to depend on the probability that the individuals of each species will procreate again after their first attempt, given the constraints imposed by their environment. If the probability is too low, the mechanisms of natural selection end up causing evolution to eliminate this second chance. In other words, if all of the individuals in a given population die after procreating once, it is because most of their ancestors managed to reproduce only once anyway.

We saw in chapter 3 that the males of certain species of insects and arachnids often die during the copulatory act. In *Latrodectus hasselti*, a species of spider related to the black widow, the male goes even further. He actually invites his partner to devour him by tipping himself into her jaws during copulation. When copulation is complete, the male is already partially digested! Who benefits from such suicidal behavior? It doesn't provide the female with a very large meal, since she is nearly fifty times bigger than the male. A third of the time, the female spurns the offering, which does not even affect the number and weight of her eggs. The currently held hypothesis is that the cannibalized male uses this technique to better ensure his paternity. Indeed, copulation lasts longer in this case, which enables him to transfer more sperm, and the female often doesn't accept another partner after the one she devours. Since the life span of the male is limited anyway—five to ten times shorter than that of a female (even if he doesn't meet any)—his sacrifice allows him to take best advantage of what is almost surely his only opportunity to reproduce.

Whenever death does provide a direct advantage, it is certainly not to the species as a whole, but rather to the deceased's larger number of heirs, sometimes by facilitating their lives. The tachigalia is yet another example of this. This tropical tree, by dying suddenly after reproduction, creates an opening in the very dense forest canopy. It thereby provides its young descendents with access to light, in some sense giving them the gift of life a second time. The young trees also have a good chance of inheriting the parent's colony of symbiotic ants. The death of the adult tachigalia thereby directly benefits its own genes, present in its offspring. The sacrifice of the male spider, or that of the tachigalia, which abandons its place in the sun and its ants to its descendants, is not a goal in itself. As Williams stressed, only an in-depth case-by-case study can define "the desired goal," if indeed there is one.

The Young Come First

In an attempt to better account for the ineffectiveness of natural selection with respect to mutations that affect old individuals, Williams put forward an additional hypothesis. According to this so-called "antagonistic pleiotropy hypothesis," certain genes or certain versions of a gene may confer an advantage at one period of life and a disadvantage at another. Metaphorically speaking, natural selection weighs the pros and cons. But it gives the pros a higher coefficient, proportionate to the reproductive prospects of the young individual who benefits. Too bad, then, for the old individual that it may one day become! It is precisely this "may" that explains why natural selection deals differently with the two effects of the gene, and therefore keeps a gene that is beneficial to a young individual even if it turns out to be harmful when the individual gets older.

How can a mutation have a different effect in the same organism when it is young and then when it is old? Young and old organisms are different in how they function, in their needs, and in the constraints that affect them. The effect of a mutation also depends on the context in which it is expressed. This context changes throughout development—as the cells divide, differentiate, move, and interact to form adult tissue. Williams's example, though pure speculation, was that of a gene involved in calcium binding. During the growth period, effective binding would be an advantage to form the bony skeleton, but would then work against the organism as it advanced in age, for example by accelerating the calcification of arterial tissue.

All other things being equal, we can predict that species that reach adult age the fastest should also age the fastest, once they reach adulthood. Indeed, their genes are quickly in a context (an adult organism that has already procreated) that is different from that for which they were primarily selected (a developing organism that must reach sexual maturity and procreate, despite the dangers lying in wait). The late-onset harmful effects of these genes may thus appear and accumulate more quickly than if natural selection had had to prevent them for a longer development and maturation period. Conversely, slowing or halting development should prevent aging by preserving an organism that is considered young by natural selection, and in whom the late-onset harmful effects of these genes do not appear. This situation is illustrated by the survival abilities of many insect species in the larval stage. Even under normal conditions, it is not infrequent for larval life to last longer than adult life. (Recall the seventeen years spent underground that precede the emergence of the adult periodical cicadas!) In times of scarcity, certain insects can remain in the larval stage much longer than usual, sometimes with dozens of additional molts.

When several alternative development pathways are programmed into the genome of an organism, each of them may confer a different life span. The most spectacular examples include social insects such as bees and ants. Depending on what food it receives in the first few days of its life, the same larva can become a queen or worker, and experience completely different fates, including as regards life span: for workers, a few months, or at most a few years for certain of the luckier species of ants; for a queen, up to twenty-eight years! The appearance during evolution of slowed aging in queens also seems related to the fact that they are highly protected, which reduces their extrinsic mortality to almost zero. What's more, this reduced rate of aging seems to be most marked in species whose lifestyle protects the queens most effectively.[22] We still do not know exactly what processes underlie the divergent destinies of queens and workers. It is also difficult to figure out which effects are caused by genes and which are caused by the very different treatment queens and workers receive throughout their existence. However, this example shows that the same genome can confer very different life spans, depending on the environment, and without involving any pathological differences. In chapter 8, we will see that the nematode has a long-life genetic program that it activates to get through difficult periods in the larval stage, but that it generally refrains from taking advantage of in the adult stage!

From the Disposable Soma to Selfish Genes

Several authors have built on the ideas presented above, or have looked at them in a slightly different light. In the 1970s, Kirkwood developed a very Weismannian theme with his term "disposable soma." Indeed, Weismann had already suspected that, under the influence of natural selection, the final step would be to sacrifice the individual for the benefit of its own lineage (but not the entire species). He even spoke of a "renunciation of the somatic cells" in favor of the reproductive cells. As Kirkwood stressed even more strongly, the only biological purpose of the organism is to ensure the continuity of the germ line from which it derives. How effectively it accomplishes this task is the main criterion for natural selection. The evolution of a life cycle, over the generations, reflects a permanent arbitration between requirements for reproduction and survival. In addition, on the time scale of an individual life, any organism is constantly being faced with a similar, crucial choice: Should it devote its resources in energy and raw materials to its own survival, or to the survival of its germ line by procreating? The question arises even if resources are plentiful and only becomes more complicated in times of famine. Marsupial mice, salmon, spiders, and many other species show that no sacrifice is too great, provided it optimizes the overall reproductive performance of the individual making the sacrifice.

[22] L. Keller and M. Genoud, "Extraordinary Lifespans in Ants: A Test of Evolutionary Theories of Ageing," *Nature* 389 (1997): 958–960.

This bargaining or bartering system has been suggested by experiments on caloric restriction. The theory is simple. It consists of reducing the caloric intake of a group of laboratory animals to roughly half of what they would eat if they were fed *ad libitum*.[23] Their life span and fecundity are then compared with those of control animals, raised simultaneously in the presence of a plentiful food supply. The general conclusion of such studies is that famine makes the animals less fertile, or even sterile, but it increases their longevity. In some cases, they partially compensate for their lower initial fertility by procreating well into old age. Since there are insufficient energy resources to ensure effective reproduction, the animals appear to invest more in the preservation of their somatic cells. This strategy gives them a better chance of holding out until times get better. This interpretation fits in well with the disposable soma theory.[24]

Instead of wondering why we die, perhaps we should be surprised that we live so long—much longer than what seems strictly necessary to transmit our genes. Leonard Hayflick insisted on the safety margin that must be provided by the soma's ability to repair itself. For each individual to have a good chance of reaching maturity and transmitting its genes to the next generation, it's better to provide it with a little more than the bare minimum, give it backup systems that it may not need. A side effect of these systems—again making a distinction from their actual purpose—is that the individual may live a little longer. The organism continues on its way, benefiting from the extra momentum provided by natural selection to ensure it a non-negligible probability of fulfilling its mission, which is to raise its progeny until they can fend for themselves. Hayflick proposed a mechanical analogy, which therefore has its faults, as we have seen, but which is telling nonetheless. A spaceship[25] is not designed to be perfect but rather to achieve certain goals. What happens afterward doesn't matter to the engineers. They are not seeking perfection, but only the best compromise possible, notably in terms of cost. They know full well, Kirkwood would add, that even a theoretically eternal spaceship will some day encounter a meteorite, a violent stellar wind, or a black hole that will destroy it. All the efforts to make it eternal will then have been for naught. However, to ensure that it has an acceptable probability of reaching its ob-

---

[23] While still ensuring an appropriate intake of vitamins and other essential elements.

[24] There are other possibilities for explaining the mechanisms involved, which we will discuss in chapter 5.

[25] This remark also applies to analogies and metaphors.

jectives, the engineers will, within reason, place the bar higher than the strict minimum. In all likelihood, the spaceship will therefore exceed its objectives. What's more, no one will be able to predict which of its elements will fail first, since in the quest for an optimal, but not necessarily perfect, design each component will have been given the same level of reliability, sufficient for the whole unit to have a good chance of success.

Without question, the most thought-provoking and stimulating discussion of the links between genes and individuals, between germ and soma, came from British zoologist Richard Dawkins, who coined the term "selfish genes." According to Dawkins, living organisms are temporary "vehicles" or "survival machines" for the genes as they travel through the generations. The selfishness of the genes is not conscious in any way, of course. It results from the way natural selection works. A gene is potentially immortal, to the extent that it can ensure the transmission of copies of itself from generation to generation. In practice, to be able to travel far into the future, genes, or gene assemblies, must define, or as Dawkins says, "build themselves" the best survival machines possible. That is where their selfishness resides, in the end. They don't care what happens to the vehicle (recall Hayflick's spaceship) once it has completed its mission.

We can easily see, here again, that Weismann had shown the way when he spoke of the body as "a secondary appendage of the real bearer of life, the reproductive cells," which are "virtually immortal." Dawkins's ideas obviously do not derive their originality only from the replacement of the words "reproductive cells" with "genes," which had not yet been discovered in Weismann's period. He took the logic that had already been expressed by Medawar, Williams, Kirkwood, and all contemporary neo-Darwinists one step further by stating that natural selection acts primarily, if not exclusively, on genes.

From Theory to Practice: Evolution in the Laboratory

What is the relevance of evolutionary theories to aging and longevity, how closely do their predictions correspond to reality? If we really want to test evolutionary theories and identify where they may be limited, observations must be completed by experimental approaches. The verb *completed* is important here, since these approaches also have their intrinsic shortcomings. They require that work be conducted in a controlled environment, which is therefore more or less artificial.

As we have seen, evolutionary theories say that there is no active, deliberate programming of aging and natural death in multicellular organisms. We can, however, say that these unpleasant processes are programmed into the genes—provided we add that it is not for their specific virtues but as inevitable side effects of the way natural selection works. Natural selection becomes less and less interested in individuals as they age, because there are fewer and fewer of them and because their relative contribution to the following generation is consequently smaller and smaller. Who cares, in the end, whether aging and death are written into the genes, as long as they don't show up too early!

That is the common basis for evolutionary theories, which then split into two main alternative versions, passive and active. According to the first version, natural selection is only passive, or powerless, when faced with mutations that disadvantage old individuals, because their effects are felt outside the period "protected" by natural selection. In more technical terms, this is known as the mutation accumulation hypothesis. This view holds that senescence and death are nothing more than an inevitable failure of natural selection. They appear despite natural selection, which normally prevents or limits the spread of harmful mutations within a given species but is much less effective when the impact of the mutations appears later in life.[26] This is why the mutational load of a species weighs more heavily on the oldest members, and eventually ends up crushing them.

The second possibility is that natural selection actively helps determine the longevity of a species, but in an indirect manner. How? By favoring a given characteristic that gives the organisms in question an advantage (reproductive, for example) even if it is to the detriment of their life span. In technical terms, this is called the life history optimization hypothesis. This view holds that aging and death result from an arbitration carried out by natural selection, which does the best it can with the resources at hand. According to proponents of the mutation accumulation hypothesis, natu-

---

[26] We can liken this notion of aging to the neutralist vision of molecular evolution. It basically says that most genetic differences that have accumulated between the species do not have a selective advantage, but that they result only from random drifts within finite populations (see P.-H. Gouyon et al., op. cit.). Here, the absence of advantage is compounded by harmful effects on older individuals.

ral selection simply prevents the worst, but according to life history optimization partisans, it does more, provided that the organism comes out
on top in terms of reproductive success. Williams's proposal—genes beneficial to the young but harmful to the old—falls within the category of this
second hypothesis, as does Kirkwood's disposable soma theory.

Life history optimization seems to be the most attractive view, for non-
scientific reasons, first of all:[27] if death is not programmed directly into the
living (if it does not have a clearly identified function), it could at least
have an indirect utility, as a "price to pay" for certain advantages. But the
life history optimization hypothesis also seems more general. Mutation
accumulation could be considered a sort of borderline type of optimization where the compensatory advantage (which benefits young individuals) approaches zero. Plus, the weight of the mutational load, in the strict
sense, results from an equilibrium, which is a form of optimization implemented depending on the environment of each species. This equilibrium
determines how far natural selection can go to counteract the load, for example defining the minimum period during which it will tend to exclude
the harmful effects of mutations.

In favor of the optimization hypothesis, we now know that there are
numerous genes (pleiotropes) that have multiple functions, through this
was not known when the theory was formulated. One isolated but very
spectacular example is provided by the components of the crystalline
lens, present in the eye of vertebrates and certain invertebrates. Biologists
have discovered, to their great surprise, that several proteins that form the
crystalline lens also play a role in every cell of the organism, and that this
role has no direct relation to vision.[28] The evolution of the corresponding
gene is thus subjected to a double constraint. Thus, there is a potential
source of conflict, since a mutation that improves the effectiveness of the
protein as a component of the crystalline lens may have an opposite effect
on its other function, and vice versa. Several major age-related human
pathologies seem to reflect this type of antagonism, in particular non-
insulin-dependent diabetes and prostate tumors. The prevalence of
Alzheimer's disease in the elderly suggests that it also may be the dark
face of genes selected for their benefits to young individuals. Similarly, a
gene that protects "younger" people from myocardial infarction seems to

---

[27] Which scientists are not the last to be sensitive to, as we have already seen . . .
[28] Specifically, enzymes involved in glycolysis, the first phase in the transformation of
glucose into energy in almost all living cells.

have an unfavorable effect later in life, but we don't yet know how it operates.[29] In both of these cases we can only speculate, since the area of human pathology hardly lends itself to experimentation, especially on genes and the evolution of the living. Researchers prefer to turn to species such as the fly, whose generations pass much more quickly.

The fruit fly (scientific name *Drosophila*)[30] currently plays an important role in research into the genetic basis for aging. This may seem natural, given the fondness of geneticists for this organism. The fruit fly did, after all, contribute significantly to the establishment of the very concept of genes, chromosomes, and mutations. Its life span was studied in the 1920s, but it wasn't until the 1980s that several laboratories began selecting for lines of *Drosophila* with altered life cycles, and specifically with increased life span. For these lengthy experiments, fly populations had to be followed for up to one hundred generations, and even with each generation lasting only ten days or so, this represented several years.

The basic principle for this selection is to allow the flies to reproduce only when they have reached an advanced age (i.e., several weeks). These flies contribute to the next generation due to the vigor they have conserved up to that point. Such an experimental protocol tends to select for individuals that age the least quickly, from the standpoint of reproductive performance. The principle is similar to the one used by breeders to obtain cows that produce more milk, more vigorous bulls, or faster racehorses. It's simply a matter of taking advantage of the natural genetic diversity of a population and favoring the desired combinations by applying the right selection to each generation. In the case of the fruit fly, in order to reach more solid and accurate conclusions, several fly populations are subjected separately to the same selection, while control populations are raised with no selection or subjected to an opposite selection. What do we see as the generations pass? For most of the laboratories that have attempted this long-haul experiment, the results seem to be similar. Life expectancy increases in the populations selected for their ability to reproduce later in life. Average and maximum longevity are approximately 30 percent higher than in the original flies, after fifteen to twenty generations. Only one laboratory, out of half a dozen, did not observe this increase, which led to an exchange of fairly curt articles in specialized journals. But this

[29] We will discuss this in more detail in the next chapter.
[30] More specifically, the species *Drosophila melanogaster*.

absence of unanimity is really not all that surprising, given the number of parameters that may affect fly breeding experiments conducted over such long periods.

One point now seems very widely accepted: longevity depends in part on genes, and certain combinations of pre-existing genes in a heterogeneous population confer longer life spans. These experiments, although protracted, are too short to allow for the appearance and diffusion of new mutations. The selection applied by researchers simply sorts versions of genes (alleles) that already exist. Certain alleles are gradually eliminated from the population whereas others, on the contrary, end up being carried by all the individuals.

Why aren't these "long-life alleles" more widely present before the experimenter comes along to artificially favor old flies? This issue is at the heart of the evolutionary question. One of the two hypotheses discussed above answers it by saying that the experimenter gradually lightens the mutational load by selecting those individuals with the least load at each generation. This purges the population, little by little, of mutations whose harmful effects appear later in life. This theory predicts that chance plays an important role in determining which mutations will be lost first. The causes of longer life could therefore differ from one population to another, even if they are subjected to an identical selection. This observation would also be compatible with the optimization hypothesis, but optimization makes an additional, distinct prediction: selecting for genes that are beneficial to the old can only be done to the detriment of the young. The selection of "long-life alleles" must have negative consequences on the first phases of life.

At first glance, these results seem favorable to proponents of the optimization theory. Late fertility and longer life are often accompanied by negative effects on early fecundity, and sometimes also on viability in the larval stage. However, the data are hard to interpret. No one can rule out the possibility that other aspects of the life cycle, for example, egg development time or sexual precocity in females, may have also been involuntarily subjected to selection. Some of the changes observed in the long-lived lines may not be directly related to their increased longevity. Moreover, the selection process does not really operate on longevity but rather on late fecundity, even if this presupposes relatively long survival. It is possible that fecundity in the young is not lost in exchange for increased longevity but rather for increased fecundity in older flies.

To respond to these objections, in October 1991, a Dutch team began a selection operating more directly on the longevity of *Drosophila*. It is obviously impossible to pair the individuals who will live the longest in each generation, since their life span can only be known after the fact. However, if we could "conserve" certain members of a sibling group while measuring the longevity of the others, we could identify the long-lived families by observing the second group, and then pair the members of the first group to obtain the next generation. This family-based selection is easier in flies, because they age much more slowly when kept cool. Their life span measured at 29°C is roughly two months maximum. At that age, their siblings kept at 15°C are still young, since the *average* life span at 15°C is more than twice as long as the *maximum* life span at 29°C! So we can put a family "in the cooler" at 15°C while we measure the life span of certain family members at 29°C.

To select a line with increased longevity, called L, researchers pair the cooled representatives of the six longest-lived families, out of the thirty families tested at each generation. The same process applied at the other end of the spectrum selects a line with a shorter life span, called S. Two independent populations are subjected in parallel to both of these selection systems in order to evaluate reproducibility. This approach has its disadvantages. It is both complex and time consuming, since the interval between generations increases from ten days to around two months. But it does, in principle, make it possible to select exclusively for the "life span" criterion. According to the initial results, which cover six generations, the two L lines gain more than 20 percent in life span. The loss for the two S lines is less marked and is significant only for male flies. One important finding is that the reproductive output of the females derived from the L lines is nearly two times lower than in the control population. The increase in longevity seems, therefore, to be achieved to the detriment of reproduction, measured throughout the entire life of the fly. The authors conclude in favor of Kirkwood's disposable soma theory, one of the main versions of the life history optimization hypothesis.

However, other arguments support the mutation accumulation hypothesis. For example, the long-lived flies obtained by various laboratories are often more resistant to different so-called "stressful" situations such as desiccation, famine, or ethanol fumes. When selection is released, the fly population gradually loses its longevity but not all of its resistance. And it is highly likely that in the initial population, low resistance and a short life went hand in hand. Selection thus "purged" a set of harmful mutations from these lines. The flies that were returned to a more "natural" selection probably did not age for exactly the same reasons as their ancestors that preceded the experiment. This suggests that senescence results from an accumulation of diverse and partly random fragilities, during the evolution of a given species.[31] It is safe to say that both the passive and active views of the evolution of aging are partly true, and it is probably pointless to endeavor to decide which is actually right. For each particular aspect of aging, however, the question remains: Does it reflect the accumulation within the species of genes with late-onset harmful effects, or is it the consequence of an advantage that benefits the young?

Conclusion: Does Death Have a Biological Purpose?

The ideas that we have just discussed shed a paradoxical light on the notion of a possible role or purpose of death. It would evidently be reassuring, or at least a consolation, to be able to find a scientific justification for death—to show that it provides an advantage, whether it be to the individual or the species. Metchnikoff even went so far as to assert that natural death corresponds to an instinct as strong as the sexual drive. He added that science should someday make it possible for everyone to live long enough for the instinct for natural death to materialize, thereby eliminating the fear of death and, by the same token, the need to resort to religion.[32] In a less ideological vein, many people have affirmed that natural selection favors the appearance of a specific death mechanism, because it confers an advantage to the group, or to the species. The thinking is that death eliminates the old, supposedly more worn out and less effective individuals, in order to keep more resources for the others. This hypothesis has always been quite popular, despite the fact that it is anthropocentric and circular, since it presupposes the very phenomenon it purports to explain—namely, the decrepitude of the old.

---

[31] Note that these are not weaknesses accumulated by aging individuals, but hereditary characteristics accumulated by the species over evolution, that determine the type of aging and the longevity of individuals of this species.

[32] See chapter 1.

The arguments presented in this chapter run counter to such thinking and show that such a theory contradicts what is known about evolutionary processes. We present one last example to illustrate this point. Most adult mammals have solid teeth thanks to the enamel that covers them, but the teeth are not replaced. In the young, these teeth are an advantage, compared with less solid teeth that might be able to grow throughout life. However, the progressive wearing of the teeth is a handicap in the oldest individuals, to the point that it inescapably limits their life expectancies. This limitation is an effect of natural selection, since it favors enameled teeth. However, it cannot be said that a reduced life expectancy has a selective value in and of itself. Natural selection simply ensured that individuals had teeth that would last long enough for most individuals to succumb to other hazards—after having transmitted their genes. R. Dawkins expressed this point by an apocryphal story.[33] Henry Ford wanted to find out if there were any parts of his famous Model T that never wore out. His people found only one, still in perfect condition in all the old cars in the junkyard. Economic logic dictated that Ford reduce the quality of this part to bring its life span into line with that of the entire car. He was able to decrease production costs without detriment to the longevity of his product!

Similarly, the same logic would require an improvement in a poor-quality component that limits the success of the product. Thus, when an increase in life span can represent an evolutionary advantage—and the resistance of the enamel over time has reached its limits—natural selection is capable of inventing new solutions. This is true for elephants and manatees, which have a system of continuous tooth replacement. Doubtless these species were sufficiently protected by their size and environment for delayed aging to be worthwhile. In any case, there is a point beyond which it is no longer advantageous to prolong life, in evolutionary terms.

Aging and the death that it precipitates must certainly be considered consequences of evolutionary processes that appeared fairly late in the history of life. They are not intrinsic properties of the living. Natural selection did not actively promote the appearance of death; on the contrary, it seems to be more involved in damage control. Unfortunately—from a

[33] In R. Dawkins, *River out of Eden: A Darwinian View of Life* (New York: Basic Books, 1995).

human point of view—its action itself has natural limits. In the logic of evolution, natural, age-related death is not a primary necessity; it has no utility of its own, in the sense that it is not a driving force in evolutionary processes. Death represents a failure of natural selection, at least as much as it represents a product of the same process.

IN SEARCH OF THE FUNDAMENTAL MECHANISMS
Genes, Cells, Humans, and Death

What are the biological mechanisms involved in the aging and death of individuals? Aging is defined, formally, as an increase in mortality rate with age.[1] This definition is statistical in nature, since it is based implicitly on the observation of a population of individuals. The link between a definition that is essentially statistical and a definition of aging for an individual is so common that it easily goes unnoticed, but it should be clarified. What is simply statistical observation at the population level becomes a probability at the individual level: namely, the growing probability of dying during a given time, as we get older. Statistically, a human being has roughly twice as great a chance of dying at age fifty-eight as at fifty, and this probability doubles again by age sixty-seven. But can we pass from this statistical view to an analysis of the specific biochemical mechanisms involved? Can we track aging down to the very cells of each individual, or to that individual's genes? This question opens the way for an experimental approach to aging. We will see in this chapter that there are myriad theories, sometimes contradictory, sometimes complementary. Let's start by saying that these theories fall into three categories: stochastic theories, programmed theories, and those that associate random chance and necessity. For example, the free radical theory, still very much in vogue after almost forty years, falls within the category of stochastic theo-

---

[1] See chapter 2.

ries. Oxygenated free radicals, fragments produced by the splitting of certain molecules on contact with oxygen, are intermediaries or natural by-products of the oxygen metabolism that takes place within a cell. They are particularly unstable and tend to react with any other molecules in their immediate environment. Free radicals can cause irreversible damage to DNA, enzyme systems, or cell membranes, thereby altering the genome of the cell and disrupting its metabolic function.

The theory known as "error catastrophe" also falls into the stochastic theory category. First proposed by Orgel in 1963 to explain aging, it has long fascinated the scientific community. This theory holds that key enzymes involved in DNA replication, transcription into RNA, and translation of RNA into protein make mistakes, such that each cell division creates cells that synthesize slightly mutated enzymes. These enzymes are likely to make more mistakes, sweeping the cells into an exponential spiral of mutation accumulation until the final outcome is that they are unable to continue dividing. The discovery in the 1970s that inactive enzymes accumulate with aging in cells or in certain aging organisms seemed to provide experimental support to this attractive theory. However, biochemical analyses have shown in all cases that the modifications observed were not mutations in the enzyme sequences, as the theory predicted, but that they were more subtle modifications, such as changes in the spatial conformation of molecules. DNA replication can work equally well in "aged" cultures as in "young" cultures. This theory did not withstand the test of facts.

Partisans of "programmed aging" theories argue that aging prolongs embryonic development and therefore is part of an overall plan whose major phases are marked by transitions controlled by specific genes. Others prefer to describe the transitions as passages of a thermodynamic system far from its state of equilibrium toward lesser levels of entropy production, joining the studies of the Brussels school on the thermodynamics of irreversible processes.[2] In this theoretical context, aging appears as the stochastic but nonetheless inevitable progression of individuals toward the ultimate state of equilibrium: death and dissolution into the environment. The question then is what controls the progress along this slope, the

---

[2] Initiated by the physicist Ilya Prigogine (Nobel laureate in chemistry, 1977).

speed of transition toward states that are closer and closer to death? For we now know that natural selection can give living organisms the means to escape death for a comparatively long time.

If we use the analogy of the genome as the life program for organisms, we must choose between two interpretations of aging: Is the program running correctly, or is the program running down? Is the organism advancing toward death in keeping with a design set in black and white in its genes, or is it sliding inevitably toward its end, in a general framework set by the genes, but without death being explicitly programmed into its genome?

Without claiming to provide a definitive response to this question, we will examine the most interesting theories at hand. Some of them have been tested experimentally, and we will look at how "biogerontologists" have managed to artificially increase the life span of certain animal species by applying these theories. We will also describe how genetic studies conducted in centenarians—or, at the other end of the longevity spectrum, in people suffering from forms of accelerated aging—have helped lay the foundations for the genetics of aging in humans.

Aging and Cells: Weismann's Theory, Carrel's Error

Since the works of Mathias Schleiden and Theodor Schwann in the late 1830s, the cell has been recognized as the basic unit of life. Every living being is composed of cells, and the cell is the smallest autonomous living entity. It is therefore natural to ask whether the phenomena of aging exist in isolated cells. If the answer is yes, is there a link between cellular aging and the aging of the organism?

The German researcher August Weismann, whom we encountered in previous chapters, suggested the first in-depth theory on individual aging based on cell physiology in 1881. Of course, this theory was based exclusively on reasoning, because the experimental techniques required to support or verify it did not exist during Weismann's lifetime. It wasn't until eighty years later, in 1961, that the first experimental results at least partially confirmed the theory. Weismann's construction is based on the observation that "death occurs because the worn tissues cannot renew themselves indefinitely."[3] Organs renew by cell division, a mother cell giving birth to two daughter cells, which in turn divide, and so forth. Since a worn or defective cell is replaced by cell division, we should attribute death not to the wear of isolated cells but rather "to a limitation of their capacity to multiply." According to Weismann, natural death occurs because cells have a limited, not infinite, capacity to divide. After a certain number of generations, the cells stop dividing, leading to irreparable wear of the organs that they form.[4] From this standpoint, the death of an individual results when the organs can no longer be repaired because the cells' ability to proliferate has been exhausted. However, "death will occur . . . long before the cells have completely exhausted their ability to reproduce; slight deregulations must indeed begin to appear as worn cells are replaced more slowly and ineffectively."

Let us emphasize two fundamental predictions of this theory. The first is that the possible number of cell divisions must be fixed for each species, in order to account for the relative invariability of the maximum life span in a given species. Second, the number of cell divisions possible must be correlated to the maximum life span observed for each species: specifically, a longer life expectancy must correspond to a higher maximum number of cell divisions for the individuals of any species. The validity of the entire argument depends on experimental confirmation of these predictions.

But before Weismann's theory could be tested, we had to wait for the discovery of a technique that would revolutionize biology in the late nineteenth century, namely, cell culture. The principle consists of removing a few cells from a living organism and trying to make them survive and eventually divide outside the original organism in an artificial environment, generally a flask or a "culture dish." This is no easy task, because the cells to be cultivated must be given the nutritional elements they need to survive, a stable temperature (generally the temperature of the original organism), and the right concentrations of oxygen and carbon dioxide.

---

[3] This quote and those that follow are taken from Weismann's speech "Über die Dauer des Lebens," given before the Association of German Naturalists in September 1881. (The English translation was titled "The Duration of Life," op. cit.)

[4] Weismann did introduce a distinction between germ cells (sperm and eggs in humans), responsible for the transmission of the genome from one generation to the next, and somatic cells (all other cell types). He believed that the limited ability to divide involved only somatic cells. It is clear that germ cells must retain an unlimited capacity to divide in order to preserve the line (see chapter 4).

However, the undertaking is worthwhile because it allows one to observe, study, and manipulate living cells isolated from their complex natural environment.

The French biologist and surgeon Alexis Carrel performed cell culture experiments that would long shape research on aging.[5] Once he succeeded in developing culture conditions for heart muscle cells taken from chicken embryos, he observed that the size of the tissue sample placed in the culture increased, actually doubling within a few days after the start of the experiment. This was because the cells of the tissue were dividing. When the cells grew to the point at which they filled the flask and could no longer divide for lack of space, Carrel took a portion of the cells and placed them in a new culture. The cells resumed their division until the second flask was filled, and a second portion was then placed in a new culture, and so on. In fact, after repeating this pattern several hundred times, Carrel and his collaborators reported that from a single sample of chicken tissue they had observed cell divisions in culture for thirty-four years before voluntarily abandoning the experiment! These fascinating results spread far beyond the scientific community. At the beginning of each year, a New York newspaper published a health report on the apparently immortal cells. Science fiction authors imagined that they might escape from the laboratory and proliferate unimpeded in the city, attacking passersby.

Carrel's experiments aroused great interest among his contemporaries, but they also deeply influenced gerontologists. His results indicated that the cells placed in culture had a virtually unlimited intrinsic capacity to divide. It could thus be asserted that the cells of a complex organism were potentially immortal, in the sense that they could renew themselves indefinitely. In light of Carrel's results, aging appeared to be an effect of the organism, which cells could escape if separated from their natural environment. The organism came to be considered a prison in which the cells were prisoners sentenced to death, unless they could escape with the help of an accomplice, the experimenter. Consequently, the causes of aging

---

[5] Carrel (1873–1944) obtained the Nobel Prize in Physiology and Medicine in 1912, for the development of vessel suturing methods. His scientific merits were tarnished by the role he played at the head of the French Foundation for the Study of Human Problems, which he founded in 1941 at the request of the Vichy Government. The experiments described here were conducted in the United States.

were not looked for in cells, but elsewhere, in the factors circulating around them, such as hormones, which regulate cell interaction, or in the environment. Weismann's theory explaining aging as a limited capacity for cell division was relegated to oblivion, especially because Weismann's nationalist position at the outbreak of World War I—the year of his death—did not help either his personal popularity outside his country or the popularity of the theories he had defended.[6]

The notion of cell immortality imposed by Carrel was to become veritable dogma for several decades. Any failure to reproduce these results was systematically assumed to be a technical problem. Carrel was wrong, however, but it wasn't until the 1960s that cell immortality was finally challenged.

The Overthrow of Carrel's Dogma and the Advent of Cell Gerontology

In 1959, American biologists Hayflick and Moorhead observed that despite Carrel's theory, at that time unanimously accepted, they could not succeed in making the embryo cells they were studying divide for any extended period. In trying to figure out why, they found that the cells that died were always the oldest ones, the ones that had stayed the longest in culture. These cells stopped dividing and died after having performed a fairly constant number of around fifty divisions. This maximum number of divisions is now commonly called the Hayflick limit. This limit is a phenomenon intrinsic to each cell type and does not depend on either the medium or on any of the surrounding cells. Moreover, the number of remaining divisions depends only on the number of divisions already performed, and not on the time that has elapsed. Thus, if cells are frozen,[7] kept for several years, and then replaced in culture, the total number of divisions that they will perform does not depend on the amount of time spent at −195°C in liquid nitrogen, but will be the same as that of their nonfrozen sisters. It might be said that for frozen cells, time is suspended.

Based on these observations, Hayflick and Moorhead suggested that cells have an internal clock that regulates their total number of potential

---

[6] He renounced all scientific honors awarded to him by countries that were enemies of Germany.

[7] Which freezes their metabolism and prevents them from dividing.

divisions.[8] Despite a certain amount of initial resistance, their conclusions regarding the existence of an aging mechanism at the cellular level finally prevailed. It has now been widely verified that normal (nontumor) cells have a limited proliferative capacity. But how could Carrel have been so wrong? As we said earlier, the nutritive mixtures required for the cells to survive are complex. They are composed of certain basic elements, such as vitamins, amino acids, minerals, and hormones, but these must be rounded out with more complex media such as blood serum. Carrel added liquid media containing chicken embryo extract to his cultures. In fact, it seems that this liquid extract was not properly purified, and that it still contained living cells. So, each time he fed his cultures, Carrel was adding fresh cells. These were the cells that proliferated in each flask, not the descendants of the fibroblasts that he had initially placed in culture. Was this a gross scientific error, or fraud on the part of Carrel or his collaborators? We will probably never know for sure.[9]

Does the loss of proliferative potential in vitro[10] described by Hayflick constitute the cellular basis for organism aging, as Weismann had predicted? If so, two major predictions must be verified experimentally to be able to affirm the relevance of this correlation. First, if there is an internal mechanism capable of counting the divisions performed by a cell, there must be a correlation between the age of the organism from which cell samples are taken and the number of divisions that they can still perform in culture. In keeping with this prediction, we observe that cells taken from a young organism can generally divide longer in culture than cells taken from an old individual, and this difference increases as the age difference increases. So, while the skin cells (fibroblasts) of a newborn have a capacity of roughly sixty divisions in culture, those of an eighty-year-old can divide only around twenty times. This relation is not absolute, however, and there is a wide range of variability from one individual to another. Second, if the loss of proliferative capacity observed by Hayflick constitutes the cellular basis for aging, there must be a correlation be-

---

[8] This hypothesis concerns normal, non-cancerous cells; we will come back to this point later.

[9] See the statement of a technician in Carrel's lab quoted by L. Hayflick: Having noticed that the cells that were proliferating in the flasks could not come from the initial fibroblasts, she was supposedly told she would lose her job if she didn't "forget" what she had seen (*How and Why We Age*, 1st ed. [New York: Ballantine, 1994]). Also see R. C. Parker, *Methods of Tissue Culture* 3d ed. (New York: Hoeber, 1961), pp. 162–174.

[10] *In vitro* means "in a cell culture."

tween the proliferative capacity of cells in culture and the longevity of the organisms from which they are taken. This correlation does seem to exist, although it is not perfect. Fibroblasts taken from a chick divide twenty-five times (a chicken can live for ten years or so), whereas cells taken from Galapagos tortoises (which live for more than one hundred years) can divide up to 130 times. Finally, in patients with rare pathologies such as Werner's syndrome and progeria, features of which resemble premature aging, fibroblasts have a considerably lower capacity to divide than those of healthy individuals.

In conclusion, all these observations suggest a connection between the loss of proliferative potential in vitro, organism aging, and species longevity. However, it is true that Hayflick's experiments were conducted on cells in culture, in an environment that had little to do with the environment inside the organism. What is the capacity of such a model to describe the "real" situation? We can also ask whether the correlation observed reveals a cause-and-effect link. Is the loss of proliferative capacity a cause or a symptom of aging? This question has fueled more than three decades of controversy, to the point that we now prefer the term "proliferative senescence" to speak of the phenomenon described by Hayflick. Moreover, the ability to divide is not a property common to all cells in the body; there are more than two hundred different types of cells in an individual. By definition, cells that do not divide cannot suffer from proliferative senescence. This includes, for example, the neurons that an individual keeps practically without renewal from cradle to grave, and heart muscle cells of the sinus node, the pacemaker of the entire cardiovascular system. It is clear that these cells, which perform high-ranking functions in the hierarchy of the body, are all the more irreplaceable because they do not divide. They belong to limited populations that decline gradually and whose aging members must compensate for the cumulative deficits. Certainly, aging and death of these key cells must have a considerable impact on the organism, perhaps a more decisive impact than the loss of the proliferative capacity of cells that can divide. So, even if today we can't state that the aging of individuals is due strictly to the phenomenon of proliferative senescence discovered by Hayflick and Moorhead, these works have nevertheless had a profound influence on the very notion of aging. Because a form of aging was shown to occur at the cellular level and appeared to be determined intrinsically in each cell, the ambition of dissecting the mechanisms of aging in vitro, in a well-defined, standardized system, became legitimate. A defined experimental system opened a vast field of research and comparisons.

Hayflick and Moorhead's discovery came in the context of the fast-growing field of genetics, just a few years after the structure of DNA and the principal phases of DNA metabolism were discovered. The image of a genome at the heart of each cell, governing its activity, suggested the idea of genetically programmed aging. And starting in the 1960s, many laboratories began to try to decipher this code.

Cellular Aging: The Role of Telomeres

Because there is a limit to the number of possible divisions for a normal cell, and because this limit is determined and constant for each species, each cell "knows" how many divisions have occurred since the initial stem cell, and it can "remember" this number even after spending several years frozen. What is this internal clock that tells each cell the generation to which it belongs?

We know that all the genetic information in a cell is grouped onto chromosomes. In human cells, there are twenty-three pairs of chromosomes. At the ends of each chromosome, there are specialized fragments of DNA called telomeres. The structure of these telomeres is the same among even very different species, and consists of a chain of repeated sequences. For example, all human chromosomes (as well as those of other vertebrates) end in several hundred repetitions of the DNA sequence TTAGGG. In 1971, Russian biologist Alexei Olovnikov suggested that on each cell division a small portion of the telomeres were lost, thereby gradually shortening the chromosome ends. To the extent that telomeres do not contain any functional genes, but merely "mute" repetitions, the loss of a small piece of chromosome end is not initially too serious. But with each cell division, another small part of the telomeres is lost. As the divisions continue, the ends of the chromosomes are inevitably nibbled away.[11] Eventually the ends become so short that the cell can no longer divide, and it dies.[12]

It was not until 1990 that Harley and his collaborators showed that the length of telomeres does indeed decrease with the number of divisions in

---

[11] Bacteria don't have this problem, since their chromosomes are circular. There are no ends!

[12] This theory is also called the *marginotomy theory*.

human cells in culture. Even more striking, they established a correlation between the state of the telomeres in a given cell and the age of the organism from which it was taken. The telomere ends were longer in skin cells taken from young individuals than in those taken from old individuals. This fundamental observation has since been confirmed in all human somatic cells that divide—endothelial cells, epithelial cells, chondrocytes, lymphocytes, etc. Telomeres also play a fundamental role in chromosome stability, acting as an anchor in the nucleus membrane. When the telomeres are damaged, the chromosomes become prone to fusions and recombinations, giving rise to various unstable forms. When the telomeres of somatic cells become shorter than a certain limit, a cellular alarm is triggered and division is prevented. If cells are forced to continue despite this safety feature, they divide a few more times, then enter a crisis phase during which chromosomal anomalies are extremely likely to occur.

In contrast, telomere length does not decrease in cells that divide without limit, such as tumor cells or germ line cells. In these cells, an enzyme called telomerase is capable of elongating the ends. It "knows" which nucleic acids to add to correct an error left by incomplete replication. In fact, the very special telomere structure is a signal and a substrate for telomerase. This enzyme exists in all cells that have a nucleus, including protozoans. However, it is repressed in somatic cells. It is reactivated when a "normal" cell is transformed into a tumor cell, and it is continuously active in germ cells. Telomerase thus seems to play a fundamental role in the process of cell senescence. The absence of telomerase in most human cells is responsible for the inevitable erosion of the telomeres, which ultimately halts proliferation. And its presence in germ and tumor cells seems to be at the heart of the "immortality" of these cells. These data suggest that the expression of telomerase in normal cells would prevent cell senescence. Several teams tested this hypothesis in 1998, and it was shown that overexpression of telomerase in human retinal or skin cells allowed the cells to divide far beyond the Hayflick limit. This study reinforced the hypothesis of the role of telomeres in cell senescence.

Let's dream for a minute. If it is true that the aging of organisms is related to cell senescence, then it should be possible to eliminate aging by acting on the telomerase activity, thereby increasing life span. Certain scientists who worked on the experiments mentioned above made stupefying statements to the press, asserting that within ten years a drug acting on telomerase might be available potentially enabling people to live up to 150 years. These unrealistic announcements were subsequently tempered, but in the meantime the share price of the company that had financed the study had skyrocketed. We must also remember that doing away with cell senescence by reactivating telomerase could lead to the formation of cancers.

Incomplete telomere replication is one of the most attractive hypotheses to account for the existence within each cell of a mechanism that keeps track of the number of cell divisions. However, the picture is actually more complex, and several aspects of aging cannot be explained by this phenomenon. For example, the telomeres of mice are much longer than human telomeres and do not shorten noticeably with age, but this does not prevent mice from aging at their fast pace.[13] The aging of yeast cells also seems to be unrelated to telomere shortening. Although this process does play an incontestable role in cell senescence, it alone cannot explain how organisms age.

The birth of cloned animals has raised new questions in this context. Cloned animals develop from an oocyte from which the chromosomes have been removed and replaced with those of an adult somatic cell. The first clone, Dolly the sheep, had shorter telomeres that her "natural" siblings of the same age. This result fits in well with the preceding model. Dolly's original chromosomes, which came from an adult somatic cell, started out with a handicap compared to those of a natural oocyte (produced from the germ line, where telomerase is active). The conclusion seemed near at hand: reproduction via the germ line, in which telomere length is maintained, appeared necessary to produce an animal with normal longevity. In fact, the situation turned out to be more complex, with the birth in 1999 of cloned calves whose telomeres were longer than in normal calves of the same age.[14] Several research groups have reproduced this finding, even when deliberately starting with nuclei taken from cells cultured until they reach the stage of replicative senescence. So it seems that we still have much to learn not only about the role of telomeres in aging, but also about the regulation of telomere length during development.

[13] This may also explain why it is easier to immortalize mouse cells.
[14] R. P. Lanza et al., "Extension of Cell Life-Span and Telomere Length in Animals Cloned from Senescent Somatic Cells" *Science* 288 (2000): 665–669 (and comments, pp. 586–587).

Pitfalls in the Search for Genes Involved in Senescence

The study of genes that play a role in cell senescence and immortality is based on the use of so-called immortalized cells. These immortal cells—cells that can divide almost indefinitely in culture—do exist. They are not normal cells that follow Hayflick's law, but rather cells that have been transformed by a virus or a carcinogenic chemical, or by exposure to ultraviolet radiation. The genetic modifications carried by these cells may be a single isolated mutation, in other words, a single error on one of the three billion nucleotides contained in the human genome. This simple mutation, if it is carried on a gene that is fundamental to controlling cell division, may alone be responsible for immortalization. Sometimes the immortalization is due to manipulations involving entire fragments of chromosomes. By studying these "immortal" cells and comparing them to normal cells, the hope is, of course, to identify what causes aging in normal cells.

In the early 1980s, experiments designed to identify the genes conferring cell immortality used the "hybrid cell" experimental approach. In this method, cells are fused in order to obtain "heterokaryons," veritable laboratory phenomena that contain two nuclei within one cytoplasm and combine the genomes of the two initial cells. These hybrids are capable of dividing and proliferating. They have properties that are the result of the combined effects of the genomes of the two cell types used to create the hybrid. Immortalized cells have been fused to normal cells (with limited proliferative capacity), to see whether these hybrids were mortal or immortal. These experiments demonstrated that hybrids created from mortal and immortal cells are mortal. The simplest interpretation of this experiment is the following: (a) there is a set of genes that control senescence in normal cells and that have *lost* their activity in immortalized cells due to genome mutations; (b) if the heterokaryons are mortal, it is because of the presence of the active version of these genes coming from the mortal cell. So, according to this interpretation, cell immortality is a recessive genetic characteristic and appears as the loss of a genetic function. Continuing these experiments, researchers crossed immortal cells of different origins. In certain cases, the hybrids of these cells were mortal. The simplest interpretation of these results is that each of the two nuclei has the function required for senescence that the other is lacking. The affected genes in the initial cells were therefore different. After fifteen years of research on this experimental route, researchers have been able to define four groups of genes involved in the control of senescence.

This is a simple model, even simplistic, according to some. And it has recently been challenged. Apparently, a large proportion of the results, published in a journal of great renown, were biased by the harmful long-term effect of a substance in the culture medium. Some of the hybrids died from poisoning, and not from senescence! This example is a good illustration of the potential pitfalls of experimental systems.

In order to identify the genetic component of cell senescence, researchers also studied gene expression that changed during this process. For a given type of cell, certain genes are expressed in young cells, whereas other genes are expressed in the same cells when they are old and their proliferative capacity has been almost exhausted. This is the case for the gene *c-fos*, which is less expressed in senescent fibroblasts, and for the genes interleukin-2 in lymphocytes and interleukin-1 in endothelial cells. Other genes, on the contrary, are activated during senescence, and there are even those that announce the imminent death of the cell and bear telling names such as terminine and mortaline. It has not been proven that the expression or nonexpression of these genes causes the loss of proliferative capacity. But these modulations in themselves suggest the existence of a genetic program, which at the very least accompanies cell senescence, and may even be directly responsible for it.

Another mechanism often considered in discussions of senescence is that of methylation. In vertebrates, 1 to 10 percent of the cytosines, one of the four nucleic acids in DNA, are methylated—that is to say, modified by the addition of a methyl group, CH_3. Methylation is carried out by specialized enzymes called methylases. This process generally represses the expression of the modified gene. Thus, methylation plays an essential role in gene expression. These "housekeeper genes," which are necessary and expressed in all cells—such as detoxification enzymes—are largely demethylated, while the genes whose expression is repressed in certain tissues are strongly methylated. A notable example is the methylation and associated inactivation of one of the two X chromosomes during development of the female embryo. Methylation is transmitted with each cell division, but not perfectly. Methylation is often lost erratically during aging, both in vitro and in vivo. The result is deregulation, often resulting in an abnormally high expression of certain genes, which may contribute to aging.

Cell Senescence and Cancer

Senescent cells that lose their proliferative capacity are a mirror image of cancer cells, which proliferate totally uncontrolled. This link is illustrated by the symmetry of certain mechanisms involved in these two cases. There are genes whose specialized role is to make sure a cell does not become cancerous. These are called "anti-oncogenes," and we will talk about them in chapter 6. When a cell becomes deregulated and runs the risk of becoming a tumor cell, the anti-oncogenes act to prevent the cell from proliferating, and sometimes even eliminate the cell. It is striking to note that two of these anti-oncogenes, $p53$[15] and Rb,[16] are also expressed in senescent cells, which have reached the limit of their proliferative capacity. This suggests that the very genes that are involved in tumor control in an organism are also involved in triggering cell senescence. This may mean that the biological brake that prevents potentially cancerous cells from proliferating out of control is the same mechanism that limits the proliferation of normal cells. If this hypothesis is true, we can predict that inhibiting the action of these anti-oncogenes will interrupt the process of proliferative senescence. This is exactly what has been observed in the laboratory—namely, that senescent cells resume proliferation if the action of anti-oncogenes is blocked.[17] So it seems that we can push back the limits of cell senescence in culture by acting on the genes that inhibit cancer.

But not indefinitely. After a reprieve of twenty or so generations, these cells go into a crisis phase and die suddenly, en masse. There is in fact a second safety mechanism, and human cells are very resistant to immortalization. Immortal cells never occur spontaneously in culture from normal cells without the use of a mutagen or an immortalizing virus. Rodent cells are less refractory to immortalization, which can occur spontaneously in culture. This propensity is associated with greater genetic instability. Moreover, those mice whose cells are so apt to undergo chromosome rearrangements often die of fast-growing tumors. In humans, the overall incidence of cancers increases exponentially with age, but over a time scale fifty times greater than that of mice.

These data suggest that cell senescence is a genetic "anticancer" program,

[15] We will discuss p53 in more detail in chapter 7.
[16] Rb is the retinoblastoma gene, involved in a hereditary form of cancer.
[17] We also know that the inhibition of these two genes can promote the formation of tumor cells.

with cell senescence and death being the alternative to carcinogenesis. Extrapolated to the entire organism, aging seems to be one of the inevitable side effects of the necessary control of cell proliferation during the life of an individual. It is the price to pay for the relatively low occurrence of cancer during the first half of life,[18] the half that is most important in terms of the reproductive cycle and therefore essential for the survival of the line.

Aging and Energy Metabolism

Certain aspects of aging present remarkable similarities within one species group, even though they take place over different durations. For example, the life spans of mammals run the gamut from one to one hundred years, starting with rodents, which live one to two years, and ending with humans, who can live to 120 years. The same aging processes, with their series of associated pathologies, occur over a time frame that is one hundred times more compressed in mice than in humans. But in 1908, the German physiologist Max Rubner observed an invariant that would later prove to be particularly fruitful for studies on longevity: namely, that mammals consume during their lifetime the same total quantity of calories per unit of weight, regardless of their maximum life span. So a rat, with a maximum potential life span of four years consumes as many calories per gram during its lifetime as a hippopotamus during its fifty years of existence. Both will burn a total of about two hundred kilocalories per gram of tissue. This value is relatively constant for all mammals, with the notable exception of humans, for whom it is eight hundred kilocalories per gram. Another aspect of the same observation is the relatively constant number of heartbeats during a life, among mammalian species. A mouse heart beats the same number of times as an elephant heart, around one million beats. This value seems to be another invariant, again with the exception of the human species, in which the number of heartbeats is around three million for someone who lives to be seventy-five years old. And heart rate is related to the energy consumption we spoke of above. Blood, rich with oxygen after passing through the lungs, is pumped through the vascular system to provide the cells with the fuel they need to produce their energy-bargaining chips in the mitochondria. So there

18 In mammals.

seems to be a set, nonvarying energy reserve in mammals that each species consumes at its own pace, called its "vital energy potential." The difference between species lies not in the quantity of the reserve, but in the rate at which it is expended—a rate that researchers call the "specific metabolic rate." It can be expressed using the following equation:

Specific Metabolic Rate = Vital Energy Potential/
   Maximum Life Span

and it can also be expressed:

Maximum Life Span = Vital Energy Potential/
   Specific Metabolic Rate

This equation sums up the theory of the "rate of living" proposed by the American gerontologists Raymond and Ruth Pearl in 1928. It indicates that the maximum life span in a given species is inversely proportional to the specific metabolic rate, and it predicts that the longevity of the species in question should increase if its specific metabolic rate can be reduced.

Let's look at whether this is possible. Metabolic rate represents, as we said, the rate at which an individual expends its energy reserves. Intuitively, we can come up with two possible methods for controlling metabolic rate. The first would be to lower the individual's body temperature in order to decrease the speed of all of the biochemical reactions that govern metabolism. The second would be to decrease the rate at which metabolic fuel is supplied to the individual, in other words, to decrease the daily calorie intake.

Changes in body temperature can be imposed on cold-blooded (poikilothermic) animals whose body temperature adapts to the temperature of the environment. Temperature changes in these animals generally result in a change in life span. In 1917, for example, the Americans Loeb and Northrup showed that flies live longer in the cold. A ten-degree (°C) drop in temperature, compared with their natural environment, increases their maximum life span by a factor of 2.5. Such changes also occur in other insect species, in worms, and in fish. In warm-blooded (homeothermic) animals, the situation is very different. Their body temperature is regulated strictly to within a few tenths of a degree and cannot be changed by outside temperatures. There are, however, conditions under which body temperature drops naturally in certain homeotherms, specifically during hibernation. It is possible to artificially lengthen hibernation by maintaining the animals at a lower temperature. Animals treated in this manner live longer, in keeping with the Pearls's theory.

The second way to decrease energy metabolism, and therefore the specific metabolic rate, is to modify the diet of the animals being studied by reducing their calorie ration. The Pearls's theory predicts that a calorie reduction will lead to an increase in maximum life span. This paradoxical prediction has been verified. In the 1930s, McCay demonstrated that laboratory rats fed a diet containing the required daily intake of essential elements (such as vitamins, minerals, and proteins), but under a controlled and very restricted calorie intake, lived longer on average than rats that received normal amounts of food (fed at will). Sixty years later, Walford and Weindruch showed that a decrease in calorie intake from 120 kilocalories per week to forty kilocalories per week increased the maximum life span of rats by more than 50 percent.[19] The basic mechanisms that underlie the effect of caloric restriction on life span are not perfectly understood. One hypothesis initially proposed postulated that the low calorie intake caused a general slowing of organism growth, which could be the reason for increased longevity. However, more recent experiments showing that caloric restriction is effective even when the diet starts at the adult age contradict this hypothesis. The famous free radicals could provide a better explanation. Let's take a look.

The Role of Free Radicals and Mitochondria

The rate-of-living theory is based on the negative correlation between longevity and tissue combustion. This naturally leads us to try to understand the mechanisms of energy metabolism within cells. We will identify some of the primary "suspects" in the aging process—free radicals— while moving away from the rate-of-living theory, which does deserve credit for putting researchers on the right track.

Breathing and food provide individuals with the oxygen and carbon compounds that are required for cell survival. Within cells, these elements are converted to adenosine triphosphate, often referred to as ATP. This molecule constitutes the primary energy source for the cell, and in some

---

[19] We will discuss caloric restriction in greater detail in chapter 8.

ways is the cell's only currency for all energy transactions. ATP synthesis occurs in the mitochondria, of which there are dozens, or even hundreds, in each cell. These somewhat particular structures are endosymbiotic in origin, which is to say that they are descendents of the bacteria that probably colonized the ancestors of our cells. This invasion seems to have evolved into a perfect symbiosis, with the bacteria producing ATP by combustion for the cell, and the cell supplying everything its guests need to live and reproduce within the cell. Mitochondria have their own genome, a single DNA macromolecule, which in humans is a circle containing only 16,000 pairs of nucleotides (compared with three billion for the cell genome located in the nucleus). They even have their own nuclear genetic code. Mitochondria replicate their DNA and divide, like miniature cells, but most of the elements they need to divide are provided by the host cells.

After observing that 90 percent of the energy consumed by mammals was produced in mitochondria, Denham Harman presented a theory establishing a clear link between longevity and mitochondria, in a 1972 article titled "The Biological Clock: The Mitochondria?" The article was an outgrowth of his 1956 theory presented in the *Journal of Gerontology* entitled "A Theory Based on Free Radical and Radiation Chemistry." The arguments that he set forth are still valid today. ATP synthesis in mitochondria initiates the transfer of electrons along a series of membrane-bound multienzyme complexes until they are eventually recovered by molecular oxygen. During this transfer, chemical compounds are produced that have an unpaired electron, which makes them extremely unstable and reactive. The molecules, called free radicals, react rapidly with the molecules in their immediate environment, thereby creating irreversible damage. The result is a gradual modification of enzyme function and a change in the properties of the cell membranes. In addition, the DNA, or genetic makeup of the cell, may be damaged. It is estimated that the DNA of a typical mammalian cell suffers nearly one hundred thousand free radical attacks every day. And at any given instant, 10 percent of cell proteins carry changes due to free radicals.

The damage is the greatest in the mitochondria; since they are the source of the free radicals, they are the most directly exposed. Mitochondria are especially vulnerable because they do not have the specialized DNA repair systems that exist in a cell nucleus. So over time there will be a loss of effectiveness of the ATP production systems, which will no longer be able to provide the amounts of ATP required. The tissues that suffer most from this degeneration are those in which the cells divide infrequently or not at all and which are the greatest consumers of energy. This is the case of the most "noble" cells of the organism: the neurons, the muscle cells, and the cardiomyocytes that drive the cardiac pump without respite. These cells suffer from the progressive inactivation of their mitochondrial functions caused by the accumulated damage to their mitochondrial DNA. In the most metabolically active regions of the brain, like the cortex, the proportion of damaged mitochondrial DNA increases considerably as we age and exceeds 10 percent in individuals over the age of eighty.

Why are these tissues that are so important for the life of the individual the ones that suffer the most, while similar damage is undetectable in cells that divide, like blood or skin cells, in healthy subjects of the same age? The explanation is simple: in a tissue where the cells renew themselves regularly, a cell that becomes weak because some of its mitochondria are less productive will be overtaken by its fellow cells, diluted in the population that is undergoing constant renewal, and finally eliminated. Selection occurs automatically among such cells, to weed out deficient cells, and this selection in cell populations prevents the accumulation of defective mitochondria. In contrast, a cell that cannot divide must continue to perform its functions—or die. When mitochondria become weak in this type of cell, how does the cell react to compensate for the deficiency? It makes more mitochondria! Signals tell the weakened mitochondria to divide, which they do, replicating their defective genome. The cell, stressed by its environment, doubles its efforts to meet the energy demand, thereby selectively amplifying the least functional mitochondria. These mitochondria have a lower output, which in turn increases intracellular damage. This is why muscle fibers and neurons become laden with deficient mitochondria as they age.

So, in cells that do not divide, selection preferentially amplifies the mitochondrial genomes that carry harmful mutations. How could such a reprehensible mechanism have evolved? Well, because it isn't reprehensible at all—at least in a thirty-year-old, in perfect health, able to produce children. If he has to make an occasional effort, run a race or solve a problem, for example, his nerve or muscle cells must adapt to an increased energy demand. This is no big deal—the cells increase their population of mitochondria, which are all perfectly healthy. This is an adaptive response to effort; all the mitochondria are called on equally. However, this mechanism did not take into account one regrettable possibility: that certain mitochondria might be defective. Too bad if a cellular compensation mechanism that works well in a young organism becomes harmful in an old organism!

The theory of free radicals just described has had considerable success, for several reasons. The first reason is aesthetic, since in this theory oxygen, which is the very life force, also produces the seeds of aging and death during the metabolic process. The second reason is the effect on life span of anti–free radical substances, the famous antioxidants. Many experimental arguments support this theory. The most notable is that the damage due to free radicals does indeed increase with age. Moreover, this damage accumulation increases exponentially, just like the mortality rate.[20]

Another major prediction of the free radical theory is that if the concentration of harmful agents can be reduced in laboratory animals, their life spans should increase. The Sohal group in Dallas "made" transgenic flies whose genomes had been modified so that the concentration of two enzymes, superoxide dismutase and catalase, was higher than in wild-type animals. These enzymes are capable of eliminating certain free radicals. Superoxide dismutase acts directly on the ion superoxide produced during the transfer of electrons along the respiratory chain, and converts it into hydrogen peroxide, a compound that is less immediately dangerous, and which is in turn processed by catalase. The transgenic flies overexpressing superoxide dismutase and catalase have a life span that is one third longer than their normal siblings! Perhaps even more striking are the results obtained on nematodes, another invertebrate model that we will discuss again in chapters 6 and 8. Nematode longevity was increased almost by half by adding chemicals that mimic the action of superoxide dismutase[21] to its normal diet. Thus, as the theory predicted, decreasing the concentration of free radicals does increase the life span of the animals treated.

How can these findings be extrapolated to mammals and humans? Would a simple treatment based on superoxidase dismutase and catalase increase our life span? Right now, there is no proof that it would, and we

[20] See chapter 2.
[21] S. Melov et al., "Extension of Life-Span with Superoxide Dismutase/Catalase Mimetics," *Science* 289 (2000): 1567–1569.

must be careful to avoid jumping to such simplistic conclusions. First, it must be established how Sohal's modified flies would behave outside the artificial, protected laboratory environment. Second, we must also realize that each of these enzymes participates in a series of interdependent reactions. As a result, the action of these enzymes, the consequence of constraints fashioned by millions of years of coevolution, cannot change significantly without provoking other metabolic modifications. It is therefore important to evaluate the impact of such a treatment on other characteristics of the animal, in particular its fertility, a parameter that is closely linked to longevity, as we have seen in the previous chapter.

Another element in favor of the free radical theory was obtained by Michael Rose and other researchers who were able to select lines of fruit flies that lived longer than their siblings. It turns out that these "superflies" are also more resistant to various forms of stress, including the stress caused by free radicals, than their wild-type siblings. In particular, they experience increased superoxide dismutase activity. So, all of these findings in fruit flies reinforce the connection between free radicals and longevity. When we select long-lived fruit fly strains, we see that they have an increased resistance to free radicals. Similarly, artificially lowering the concentrations of free radicals in animals by increasing the levels of detoxification enzymes (such as superoxidase dismutase and catalase) increases their life span.

Another experimental model on aging provides interesting arguments in favor of a direct link between free radical production and longevity. The filamentous fungus *Podospora anserina* grows filaments that normally reach ten to twenty centimeters in about two weeks, from a single initial spore. The ends of the filaments then blacken and show no more signs of life. If a fragment of the filament is transplanted into a new box, it resumes growing, but the growth period is shorter the closer the filament was to the "dead" end of the original filament. This strictly determined mode of senescence calls to mind the Hayflick limit for animal cells in culture. However, it does happen (very rarely) that a filament continues to grow almost indefinitely, reaching up to 1.5 meters. These "immortal" fungi have been shown to be mutants whose mitochondrial genome is altered.[22] Their respiration and growth are slowed and the most recent findings in-

---

[22] Specifically in the laboratory of Léon Belcour, in Gif-Sur-Yvette (near Paris).

dicate that they produce fewer free radicals.[23] Longevity-mutant *Podospora* return to normal and become "mortal" again once their normal metabolic rate has been restored.[24] It has yet to be determined whether there are special signals that control senescence with no link to the production of free radicals, and especially whether the mechanisms at work are related to those involved in aging in other models such as nematodes and fruit flies.

The theory linking longevity and free radicals offers a new vision as compared to the rate-of-living theory proposed by the Pearls. They postulated that each species had a fixed energy reserve, a sort of universal constant among the living, and that the life span of an individual depended on the rate at which the reserves were depleted. What the free radical theory says is fundamentally different. Energy consumption is not *in itself* the factor that determines longevity. It is the by-products of energy consumption, the free radicals, that are so toxic that they progressively damage the organism. In this framework, longevity does not depend on an energy reserve or on the rate of energy consumption, but rather on the *rate of the associated production of free radicals*, or on the *effectiveness of systems designed to eliminate them*. There is no mysterious, universal metabolic constant silently manipulating the longevity of each species. Instead, we need to study, case by case, how organisms metabolize energy, and what evolutionary constraints helped shape their metabolic systems.

Hereditary Factors in Human Longevity

The longevity of a species—just like other characteristics, such as size, gestational period, or the age of reproductive maturity—can be considered a genetic trait typical of that species, in that it is part of the genome of the species and can be transmitted from one generation to the next. The fruit fly experiments that we described earlier show that relatively simple

---

[23] E. Dufour et al., "A Causal Link between Respiration and Senescence in *Podospora anserina*," *Proc. Nat. Acad. Sci. USA* 97 (2000): 4138–4143.

[24] S. Lorin et al., "Overexpression of the Alternative Oxidase Restores Senescence and Fertility in a Long-Lived Respiration-Deficient Mutant of *Podospora anserina*," *Molecular Microbiol.* 42 (2001): 1259–1267.

genetic changes—such as the modification of only two genes, those coding for superoxide dismutase and catalase, out of the more than ten thousand that exist in this animal—make it possible to obtain strains whose life span is substantially longer. These findings argue in favor of a simple genetic pathway determining longevity, in which just a few genes could be responsible for controlling life span. Are there genes that control life span in animal species, and if so, which genes are they? Are these mechanisms common to the animal kingdom, and what about in humans? Let's take a closer look at these questions.

In mice, cross-breeding experiments with strains that have different longevities suggest that many genes on several chromosomes are involved in controlling life span, in direct contrast to the situation we described for the fruit fly. However, Cutler and other researchers have suggested that in humans only a few genes are involved. They argue that the maximum life span increased rapidly during the evolution of the line of hominids that led to humans, which tends to rule out the possibility that a great many genes are involved in this trait (the more genes that are involved in a trait, the more slowly it evolves over time). It is nonetheless hard to know whether the short life span of cavemen was actually due to their genetic baggage. Perhaps they would have been able to reach a ripe old age had they lived in a friendlier environment.

The genetic basis of longevity in humans cannot be deduced simply by studying lower organisms. Let's recall the two major questions about longevity. First, why does a cubic centimeter of mouse age roughly one hundred times faster than a cubic centimeter of human? Second, why do certain individuals live longer than their fellow creatures of the same species? In other words, what is behind the interindividual differences in longevity within a species? The first question is a matter of comparative zoology for the "longevity" trait. But geneticists have concentrated most of their efforts on the second question, especially because the pioneers of heredity paved the way more than a century before!

Certainly, the idea that longevity runs in families has been around forever. However, this idea could not be supported without the framework of the laws of heredity. Gregor Mendel, the famous Austrian monk, crossbred different strains of peas, observed their color, texture, and size, and by using rigorous mathematical analysis, discovered the fundamental rules of genetics. In 1865, he presented his results in a paper that was largely ignored, probably because it was too unorthodox for his contemporaries. He died in 1884, and it wasn't until thirty-five years later, in 1900, that "Mendel's Laws" were rediscovered.

The twentieth century opened with an exceptionally fertile decade. Archibald Garrod described the first human genetic disease in 1902 and introduced the concept of an "inborn error of metabolism" in 1909, foreshadowing the functional definition of the gene on which modern genetics is now based. It is interesting to note how these works reflect the progress of history, in which each step paves the way for the next. The fusion of male and female nuclei had been observed in 1875. Ten years later came the description of meiosis, which is the special type of cell division by which gametes (sperm and eggs) are produced. While Hardy and Weinberg in 1908 were laying the foundations for population genetics, Sutton and Boveri proposed the chromosomal theory of genetics. A few years later, Morgan and his team showed that genes, which at that time were known to be responsible for passing on a trait, were physically located on the chromosomes. This explained why some traits are transmitted in a correlated manner from one generation to the next during meiosis, since they depend on genes residing on the same chromosome.

So the scene was set for adventurous minds to look into the hereditary nature of the trait we are interested in here, namely longevity. The twentieth century really was the century of genetics and communication, the proof being that Alexander Graham Bell himself, the inventor of the telephone, turned his attention to heredity and longevity. Bell, an avid genealogist, had access to numerous family trees, including that of William Hyde, who had no less than 22,000 male descendents and 18,000 female descendents. In an analysis published in 1918, Bell showed that the Hydes whose parents had lived to be older than eighty lived on average twenty years longer than those whose parents had died before the age of sixty.[25]

A few years later came Raymond Pearl, and his wife Ruth, who together conducted an imposing study on human longevity. The Pearls found 2,300 people over the age of ninety, including 197 over one hundred, in the New York area, and reconstructed their genealogies, going back two generations and listing the longevities of the parents and grandparents. They did the same with short-lived people who died before the age of sixty. They

[25] Quoted in L. Hayflick, op. cit.

then compared the direct ancestors of these two groups using an index invented for the purpose, which they called the "total immediate ancestral longevity index," obtained by adding the age at death of the six direct ancestors. The results of this impressive study were published in a book called *The Ancestry of the Long-lived*, which contains a wealth of numerical data. We will repeat only two items here: the "long-lived" subjects (who exceeded the canonical age of ninety) had four times more long-lived ancestors than the other subjects, and when both parents lived past the age of seventy, the chance that their offspring would reach ninety or one hundred years of age was twice that of the rest of the population.

The Pearls used what means they had at hand (direct questions and newspaper clippings) to find their "long-lived" subjects and reconstruct the genealogies without worrying about verifying their information. Forty years later, this pioneering study was repeated by a group of epidemiologists who were determined to apply the rigorous methods of their field in an area that had inspired so much fanciful conjecture. They examined the descendants of the Pearls' nonagenarians. Most of their children were dead, which made it possible to draw up survival curves. By grouping each curve according to the age at death of the nonagenarians' spouses, they were able to distinguish what influence the parents of each sex had on the longevity of their offspring. It was revealed that the propensity to longevity was indeed transmitted from mother to daughter, from mother to son, from father to son—but to a much lesser degree from father to daughter.

It is now widely established that longevity is in part hereditary, less so than the Pearls believed, but it is still very difficult to determine accurately. Many traits are genetically complex, in the sense that they result from the combined action of several genes. They are not transmitted from one generation to the next according to the simple pattern described by Mendel for the transmission of the color of his peas, a trait that depends on only one gene. When we are astonished to see the spitting image of a distant ancestor reappear in a family, with similarities that are as improbable as imponderable, it is a fluke of genetics that apparently reproduced a certain combination. The same is true for longevity, and the study of its genetic basis has been complicated considerably by its complex mode of transmission.

Watson and Crick's 1953 discovery of the double helix structure of

DNA, followed by the elucidation of the genetic code and the molecular mechanisms of gene expression, paved the way for the exploration of the molecular basis of heredity. A few basic concepts about these mechanisms are presented in the inset.

Double Helix and Gene Expression

A DNA molecule is composed of two linear polymers (also called strands) paired to one another in the form of a double helix. Each strand is composed of the sequence of basic links called nucleotides. There are four different nucleotide bases: adenine (A), cytosine (C), guanine (G), and thymine (T). A strand of DNA can therefore be represented as a specific alignment, or sequence, of As, Cs, Gs, and Ts. Each strand is complementary to the other in that an A can be paired only with a T, and a G only with a C. These two strands carry the same information, one being a sort of "negative" of the other. The separation of the strands is the basis for the transmission of the genetic information from the mother cell to the two daughter cells during cell division. Once the two strands are separated, they serve as a template for the synthesis of their complementary strand, thereby regenerating two complete double strands. In mammalian cells, DNA is contained within the nucleus in the form of chromosomes. Each chromosome is made up of a very long double helix of DNA. In humans, the length of the strands varies from one hundred million to three hundred million pairs of nucleotides. DNA is extremely compacted. If we stretched out all the DNA contained in the nucleus of any human cell, we would have a string nearly a meter long! (Whereas a cell is only a few dozen micrometers in diameter.)

Genes are distributed along chromosomes. They act as a blueprint for the synthesis of proteins, the building blocks of cells. The cellular machinery first reads the gene sequence and synthesizes RNA molecules (faithful reproductions of the initial gene), then "translates" the RNA molecules into proteins. Modifications in the sequence of a gene would cause changes in the structure and activity of the corresponding protein. Each different form of a gene is called an allele. Allelic variations may be caused by a natural polymorphism, and constitute the basis for interindividual genetic diversity: the genome of each individual is totally unique, which is why we speak of genetic print, similar to fingerprints. However, certain mutations can have pathological consequences when they result in the loss of activity in a fundamental protein, or on the contrary induce abnormal, harmful protein activity. These mutations are at the origin of hereditary conditions.

Human genetics first tested its wings on monogenic hereditary diseases. These are the simplest inherited diseases, since they result from alterations on a single gene. More than 7,000 genetic diseases have been identified among the twenty-four human chromosomes,[26] and this number is growing rapidly. The most sadly famous of these single gene diseases for which the mutated genes have been identified include cystic fibrosis, Duchenne dystrophy, infantile spinal muscular atrophy, Huntington's disease, and certain forms of early-onset Alzheimer's disease. There is still much to discover, since theoretically each gene can give rise to diseases if it is defective, and the human genome contains some 30,000 genes![27]

In the case of polygenic traits such as longevity, an initial genetic approach consists of making hypotheses regarding genes that may be involved. We will study genes that participate in functions whose preservation is vital for enabling an individual to live long. For example, we will look closely at genes that regulate the function of the immune system, the cardiovascular system, and the nervous system. We will then analyze in detail the sequence of such genes within a group of particularly long-lived individuals, and then compare it with the sequence observed in "average" individuals. The goal is to see whether particular forms of these genes (certain alleles), are more often found in long-lived individuals than in others. By using this approach, researchers have identified three genes that may play a role in the genetic control of longevity in humans.

HLA Genes

The HLA system is a key element in the immune system.[28] HLA proteins are present on the surface of all cells in the body and can be used for identification purposes, to determine whether a cell comes from a certain individual—whether it is *self* or *non-self*. The HLA system plays a fundamental role in enabling the body to recognize and destroy infectious and other foreign agents. HLA identification is also fundamental in determin-

---

[26] There are 22 non-sex chromosomes and two sex chromosomes, X and Y. Women have two X chromosomes (XX), and men have one X and one Y (XY).

[27] We know of several hundred mutations in certain human genes that give rise to diseases that differ depending on the type of mutation within the gene; this increases the number and variety of genetic diseases.

[28] HLA, "human leukocyte antigens." Jean Dausset won the Nobel Prize in 1980 for his work on HLA.

ing transplant organ compatibility for a recipient, in order to prevent a transplant rejection.

In the 1980s, several teams demonstrated that the HLA genes presented a particular profile among centenarians. Certain alleles on these genes were more often represented in these subjects than in "average" individuals. The existence of a correlation between longevity and the HLA system highlights the importance of the immune system in the aging process. This observation reinforces the intuitive notion that defense mechanisms play a central role in preserving an organism over time. Other elements support the hypothesis that the immune system is one of the keys to aging—*the* key according to certain authors. As people age, there is an increase in mortality due to infectious diseases, which account for half of the causes of death over the age of sixty-five. There is also an increase in the appearance of autoimmune diseases such as erythematous lupus and certain types of arthritis. These conditions are due to a deregulation of the immune system, causing it to attack tissues that are "self" as if they were foreign elements. We also know that there is an increase in the incidence of cancers with age (55 percent of cancers occur in people over the age of sixty-five). This can be interpreted as an inability of the immune system to recognize and destroy tumor cells in older individuals. Finally, the immune system in the oldest subjects is also less effective at producing antibodies after inoculations.

So the immune system of older subjects appears to be particularly fragile. It is less effective at recognizing and destroying invaders, and it has an increased tendency to attack self tissues as if they were non-self. It is tempting to speculate that the forms of HLA genes identified in centenarians may limit and delay this type of malfunction, and thus offer individuals that carry them more lasting and effective immune protection.

Apolipoprotein E

In 1994, François Schächter and his collaborators, in Paris, reported that certain specific forms of the gene apolipoprotein E were found among a group of centenarians in different quantities than in the average population. Apolipoprotein E is a compound that plays a number of roles. It coats certain fats that contain lipids and cholesterol to facilitate transport from the liver to their destination organs, where apolipoprotein E facilitates the distribution, penetration, and metabolism of the fats by the cells. Inside certain cells, apolipoprotein E may also be involved in the evolution of cell morphology by shaping the cell wall (composed of lipid derivatives). This aspect may be particularly important in the nervous system, where apolipoprotein E may play a role in neuron repair and regeneration after a lesion.

In humans, there are three apolipoprotein E alleles, numbered 2, 3, and 4. Allele 4 is particularly underrepresented in centenarians, whereas allele 2 is twice as common compared to the general population. The observation on allele 4 is not surprising, since we know that it is a risk factor for certain cardiovascular diseases, such as atherosclerosis (related to hypercholesterolemia) and Alzheimer's disease.[29] These two diseases represent the major causes of death in older subjects, which decreases the probability of finding allele 4 in people who live to be one hundred or more. And the overrepresentation of allele 2 in centenarians indicates that this allele may confer protection to individuals who carry it. Exactly how this protective effect is obtained has yet to be determined.

Angiotensin Conversion Enzyme

The same team also performed research on another gene that plays an important role in the cardiovascular system: the gene for angiotensin conversion enzyme (ACE). Behind this mysterious name hides one of the major players in regulating blood pressure. ACE inhibitors can be prescribed as part of a treatment for lowering blood pressure.

The ACE gene also presents several alleles. In humans, the two main, most common alleles are called D and I. A given individual can have two D alleles in his or her genome[30] (he or she is said to be homozygous for al-

---

[29] Be careful not to misunderstand the meaning of "risk factor." The presence of apolipoprotein E allele 4 in an individual does not in any way predict that the person will develop Alzheimer's disease. Risk factor is a term used in probability and statistics for a population. There are many ApoE4 individuals who have never developed Alzheimer's disease.

[30] Remember that we have two copies of each gene, except those carried on the X and Y chromosomes in males.

lele D, or DD), two I alleles (homozygous for allele I, or II), or one I allele and one D allele (heterozygous, DI). DD individuals have the highest levels of ACE in their blood, and II individuals have the lowest levels. As we might expect, a factor like ACE seems to play a role in cardiovascular disease. In particular, it has been demonstrated that there is a significantly higher incidence of myocardial infarction in DD individuals than in the general population. Allele D could be a risk factor for this type of pathology.

What about centenarians? Well, unexpectedly, there are more DD individuals among centenarians than in the control population. This result is surprising. In fact, it indicates that the DD combination may represent both a risk for myocardial infarction and a positive factor for longevity. In other words, DD individuals are more likely than the general population to have a heart attack when young, but if they manage to avoid one, they are more likely than the general population to live to be one hundred or more! The exact mechanism that makes the ACE D allele a risk factor early in life but then a favorable element later in life has yet to de determined.[31]

This result highlights the fact that the impact of a given gene on survival may vary considerably during a lifetime. This is doubtless a consequence of age-related changes in the physiology of an individual, which modifies the biological environment in which a given gene is expressed, the partners with which it interacts, and the structure of the system to which it belongs. Chapter 4 highlighted the theoretical importance of those genes whose effects on life span reflect an antagonism between young subjects and older individuals.

Accelerated Aging Genes

While we may be astonished by the extraordinary longevity of centenarians, there are also, at the opposite end of the spectrum, particularly cruel forms of accelerated aging, such as Werner's syndrome and progeria. These are rare hereditary diseases. Werner's syndrome affects one out of every five hundred thousand people and is transmitted recessively,

---

[31] Similarly, we do not know why II is unfavorable to longevity.

while progeria is transmitted dominantly.[32] They do not necessarily reflect the processes responsible for normal aging, but they do provide a striking caricature of certain aspects of normal aging. By the age of twenty, Werner's syndrome patients have gray hair or premature baldness; thin, wrinkled skin; and blurry vision due to cataracts. Their faces undergo the typical changes that doctors call "bird-like appearance." They "look extremely old" and are likely to suffer from cardiovascular disorders, cancer, osteoporosis, and other diseases associated with old age. Werner's patients often die before reaching the age of fifty. The life expectancy of people suffering from progeria is rarely more than twenty years.

The gene whose mutation is responsible for Werner's syndrome was identified in 1996 by the U.S. team of Gerard Schellenberg. It is a relatively complex gene that codes at least in part for a helicase, an enzyme that literally unwinds the double helix of DNA when it is to be read by the cell machinery to be transcribed into RNA. DNA must also be unwound when it has to be repaired or duplicated during cell division. The mutations found in Werner's patients likely disrupt all of these functions. One interesting hypothesis is that the helicase affected is more specifically involved in DNA repair mechanisms; the mutation then blocks this repair activity. The cell becomes incapable of repairing the damage that naturally accumulates with time in its DNA, hence the acceleration of the natural cell senescence phenomena. In fact, it has been noted that the cells of Werner's patients have a limited proliferative capacity.[33] If this hypothesis proves to be true, it would reinforce a theory that establishes a link between the accumulation of errors in somatic cell DNA and aging. It would also open the way for a potential tactic that seeks to increase longevity by acting on DNA repair mechanisms.

[32] In diseases with dominant transmission, a mutation on only one of the two alleles of the gene responsible will cause the disease to appear. In recessive disorders, both alleles must be mutant.

[33] See page 104. Chapter 8 will present the surprising, recently discovered similarities between Werner's syndrome and yeast aging.

CHAPTER SIX

PROGRAMMED DEATH THAT BENEFITS LIFE

The preceding chapter looked at cell physiology to help explain the mechanisms of aging and death in individuals. We described the phenomenon of proliferative senescence during which a cell line progressively exhausts its reproductive capacity. Whether this is age-related decrepitude or an extremely useful system enabling the organism to resist cancer is a question still up for debate. In this chapter we review a more extreme form of decrepitude, which leads the cell to a rapid death. As in the case of proliferative senescence, we would naturally tend to think that cell death must be a malfunction. Far from being a disorder, however, the cell death in question here is, paradoxically, indispensable for the body to function properly. Despite all the experimental indications suggesting the importance of cell death for the life of the individual, this reality has only recently come to be accepted by the scientific community.

The fact that certain fetal structures exist only temporarily and disappear after birth has been known since ancient times. Galen (A.D. 131–200) observed that the arterial canal, which in the fetus allows blood from the pulmonary artery to join the aorta without passing through the lungs (since the fetus does not breathe) disappears at birth.[1] He did not call this phenomena cell death—and for good reason, since the cell was at that

[1] See the excellent publication of P. G. H. Clarke and S. Clarke, "Nineteenth Century Research on Naturally Occurring Cell Death and Related Phenomena," *Anat. Embryol.* 193 (1996): 81–99.

time an unknown entity. The first works on cell death followed closely after Schleiden and Schwann formulated their cell theory in 1839. These two German biologists, working with plants and animals respectively, established the fundamental role of cells as the building blocks of all living organisms. In 1842, the German Karl Vogt published the first works to describe cell death in an animal during metamorphosis. At that time, the twenty-five-year-old Vogt was a refugee in Geneva, having fled Germany because of his political ideas.[2] With Louis Agassiz at the College of Neufchatel, he conducted a study on the development of the midwife toad (*Alytes obstetricans*). One of the objectives of this work was to use a microscope to observe the fate of various cell groups during the metamorphosis of the tadpole into a toad. The axial skeletal stem of the tadpole (the notochord) is replaced by the spinal column in the adult animal. Did the cells of the notochord transform into the cells that form the spinal column, or did they disappear to make way for those of the spinal column? Vogt showed that the second hypothesis was correct. That was not a case of the transformation of one cell type into another but rather the elimination of the cells of the notochord—in other words, cell death.

More than twenty years later, August Weismann and subsequently others described the existence of cell death in insects, again during metamorphosis. Cell death was also observed during embryonic development in species that do not undergo metamorphosis. In 1906, the French cytologist Collin described the mass disappearance of neurons of the spinal cord in chicken embryos. He showed that these neurons are initially produced in excess but then many of them are eliminated.

All of these results were fundamentally important; yet, after their initial observations, neither Vogt nor Weismann nor Collin showed further interest in cell death during their long careers. Vogt and Weismann's works had focused on animals during metamorphosis, and perhaps they felt that the cell death they had observed was a phenomenon specific to this particular type of development. However, it does seem astonishing that Weismann did not explore whether cell death had a more general role, since he speculated so long about the cellular basis of death in the context of his theory of evolution. It also seems that Collin simply did not realize the im-

---

[2] He later became famous following a controversy with Karl Marx, who accused him of being an agent provocateur in the pay of Napoleon III (pamphlet *Herr Vogt*, 1860).

portance of his discovery, since he never brought it up again. In fact, in a long monograph devoted to the development of the nervous system, published at the end of his career in 1944, he did not even mention his work on cell death.

These examples illustrate how hard it was for scientists to integrate the idea of cell death into a conceptual plan in which this phenomenon might play a normal physiological role. This tendency to overlook the physiological importance of cell death persisted much longer than Collin's work. In the 1920s, M. Ernst and A. Glücksmann, two students of the German embryologist E. Kallius, systematically described cell losses that occurred during embryogenesis in different structures in vertebrates. They discovered that cell loss was a general phenomenon that could well be the subject of a separate study altogether. However, the notion that cell death might play a physiological role—and not simply a pathological one—took some time to migrate beyond embryology circles to reach the general field of biology. Researchers whose main subject of study was precisely the cell—namely, cell biologists—until very recently showed no interest in cell death as such. The main reason for this negligence is indeed conceptual. For cell biologists, the cell is by definition *living*, a building block of a living organism. It represents a condensed version of the very essence of life. In this context, the essential role of the cell is to live and proliferate. Cell death cannot have a biological function and can only be considered pathological, or at best accidental. The death of a cell could only be due to the failure of a vital function, caused by the presence of an external aggression.

Cell death, however, is truly a fundamental phenomenon for the formation of an individual during embryogenesis. It is also a process that plays a determining role for the proper functioning of the adult organism, a role that is probably just as important as cell division. Most cells have an intrinsic ability to eliminate themselves—to commit suicide, if it is physiologically necessary for the tissue, the organ, and ultimately the individual to which they belong. For all the organs to form correctly during embryonic development and to preserve optimal properties after birth, many cells *must* die. We might even say that there is a genuine process of *programmed cell death*. As we will see, the origin of this process probably lies far back in the evolution of our species. As an example, we describe two particular cases: the emergence of fingers in the embryo and the formation of the brain. We then show the importance of this phenomenon in adults.

In fact, the discovery of certain genes that are responsible for programmed death in the nematode, a worm that is theoretically quite distant from humans, has made it possible to scrutinize the intimate mechanisms of this process. In this chapter, we attempt to elucidate the functions of programmed cell death. Why, during a lifetime, is it indispensable for the organism to create cells whose only fate is an early death? Why do cells have to die for the individual to live?

Sculpted Fingers: Programmed Cell Death Required to Shape Living Beings

During embryonic development, a structurally complex living being with diversified, specialized organs is created from the fusion of two starting cells, egg and sperm. This notion of *construction* associated with embryogenesis calls up an intuitive image of an object—a monument, a building, for which construction means adding building blocks one after the other. For a house, each story is created by piling up bricks to form the walls; the house as a whole is formed by the successive addition of several stories. This vision however, does not depict how living organisms are formed during embryogenesis. During formation, the embryo goes through periods of construction that alternate with periods of "deconstruction," or massive cell destruction. These cell elimination phases are just as necessary for the proper development of the embryo as the construction phases.

Let's look at the example of the development of the fingers. In mice, whose total gestation period is about 20 days, the limbs appear between day 9.5 and 12 of gestation. At this stage, the fingers have not yet appeared, and the ends of the limbs, which will become the hands and feet, look like stumps. At day 13, we see four nearly parallel regions of cell death at the end of each limb, where the space between the fingers will appear. This is the elimination phase of interdigital tissue, which will enable the fingers to separate and become individualized. In web-footed animals, such as ducks, this phenomenon does not take place, so the feet remain webbed (except for fingers one and two, which are separate and for which the interdigital tissue is eliminated). So the fingers are not formed by being built onto the ends of the limbs. It is more a process of sculpting, shaping, and carving, whereby the fingers are released from the stump on the end of the limb. The sculpture is created by destruction, causing the death of the interdigital cells. This programmed death plays a fundamental role: If it did not occur, the structure of the limb would be radically modified—it would be webbed.

The process by which limbs appear illustrates the importance of cell death for the formation of living organisms. Cell death here plays a morphogenetic role—that of generating the shape of the body. To return to the analogy of the house under construction, we might say that some cells that appear during embryonic life play the role of the scaffolding. Once their mission is accomplished, these cells are eliminated, revealing the final form of the organ that they helped construct. Cell death also has a morphogenetic function during the metamorphosis of insects and amphibians, a process that involves radical changes that would be impossible without the death and disappearance of entire populations of cells. This process is how the caterpillar transforms into a butterfly, and the tadpole into a frog.

Programmed Cell Death and Brain Formation

The human brain has roughly one hundred billion neurons—the nerve cells. Each of these cells establishes contacts with other neurons to exchange information in the form of electrical and chemical signals. These contacts take place at the synapses, where each neuron can establish up to 150,000 contacts, with the average being one thousand. There are an estimated 100,000 billion synapses in the human brain. This means that there are one hundred to one thousand times more synapses in the brain of a human being than there are stars in our galaxy!

For the nervous system to function properly, the neural network must be wired with great precision. How can each neuron distinguish, from among the one hundred billion other neurons, the thousand (or more) with which it will form synapses? The precise mechanisms that set up the neural network during embryonic life are not fully understood, but it is thought that neuron death plays an important role in this process. If we systematically count all the neurons present in the brain at all stages of embryonic development, we observe at different moments massive waves of neuron disappearance of a magnitude that varies from one region of the brain to another. It has been demonstrated that in certain brain regions, more than 80 percent of the neurons created during the early phases of embryonic life disappear before birth![3]

The "neurotrophic hypothesis" has provided a useful conceptual framework for understanding these massive eliminations of neurons. This theory was based on the works carried out by Rita Levi-Montalcini and Viktor Hamburger in the 1940s, and then during the subsequent twenty years by Rita Levi-Montalcini and Stanley Cohen, who won the Nobel prize in 1986 for their discovery of "nerve growth factor," or NGF. According to the neurotrophic hypothesis, neurons are produced in excess during embryonic development. The neurons need an essential growth factor to survive, which is provided only by the target that they must reach. This essential factor, called growth factor, varies from one neuronal population to another, which could explain how developing neurons are guided toward their targets. Also according to this hypothesis, the target secretes a limited amount of growth factor. The neurons that converge on this target cannot all obtain the trophic factor. Those that capture the factor survive, and the others die. This second aspect of the neurotrophic hypothesis—competition among neurons caused by the scarcity of growth factor—accounts for the neuronal death observed during embryonic development.

The merit of a great theory is that it not only accounts for experimental facts but also gives rise to predictions that further clarify the systems being studied. If the neurotrophic theory is true, it will help us understand the utility of neuronal death during embryonic development. Let's look at a neuron that heads for the wrong target. According to the neurotrophic hypothesis, this neuron will not receive the trophic factor it needs, and it will therefore die. In theory, embryonic neuronal death serves to eliminate neurons that establish aberrant connections; it corrects wiring errors. Now, consider the fact that according to the neurotrophic theory, a target releases only a limited amount of trophic factor. The amount of factor released is proportional to the size of the target; consequently, the number of neurons that reach it and survive increases with the size of the target. Via this neurotrophic mechanism, neuronal death serves to fine-tune the number of neurons heading for a target to the size of the target. It adjusts the number of connections.

The general scope of the neurotrophic theory, as well as a number of its

[3] R. W. Oppenheim, "Cell Death during Development of the Nervous System," *Annual Rev. Neurosci.* 14 (1991): 453–501.

principal predictions, have been verified. In particular, the role of neuronal death during embryonic development has been shown to be one of correcting wiring errors, as has the relationship between the size of a target and the number of neurons that innervate it.[4] Many trophic factors have also been isolated and described in the nervous system. There exists a series of compounds that are capable of keeping nerve cells alive, at least during embryonic life. In the next chapter, we discover how these growth factors can be used to treat nervous system diseases, in which neurons die not to benefit the individual, but for pathological reasons.

Physiological Cell Death Also Exists in Adults

We have seen that cell death plays a determining role during embryonic development. But there is also a process of cell self-destruction in adults. This type of cell death is very different from accidental death. Accidental death is a result of sudden, toxic changes in the cell environment that the cell cannot tolerate. Death by self-destruction, in contrast, is an important adaptive process that is useful to the organism.

In 1965, the Australian J. F. Kerr demonstrated the existence of a physiological cell self-destruction process.[5] He was studying cell degeneration in the rat liver, subsequent to ligation of certain branches of the portal vein.[6] This operation reduces blood flow in the lobules of the liver supplied by the ligated vein branches. In addition to the rapid wave of cell death that occurs shortly after the lesion, another type of cell death occurs during the weeks that follow the ligation. That type of cell death, however, is slow and progressive, leading to a decrease in the size of the lobules whose blood supply has been affected, and allowing the organism to adjust the size of the lobules to the remaining blood flow.

Two different types of cell death are observed in this experiment. The first occurs within a few hours after the ligation. The second is a slow

---

[4] However, as with any theory, the neurotrophic hypothesis is only a schematic representation of reality. We now know that there is a growth factor specific to each neuronal population, which means that there are other mechanisms at work that explain the specificity of neuron-target recognition.

[5] J. F. R. Kerr, *J. Pathol. Bacteriol.* 90 (1965): 419.

[6] The portal vein carries blood from the abdominal organs to the liver. It divides into two branches, one for the right hepatic lobe, and the other for the left lobe, each branching to reach all of the lobules. Blood flow to the liver is also ensured by the hepatic artery.

death that occurs during the subsequent weeks. The cells eliminated during the first wave die "accidentally" because they cannot survive the sudden shortfall of blood flow. In contrast, the second wave of cell elimination allows the liver to adjust to the persistent deficit of blood flow. *It reestablishes the balance between blood flow and the size of the organ.* This type of death plays a role in maintaining hepatic *homeostasis.*

By looking through an electron microscope, we see that these two types of death present totally different morphological features. Initially, the cells that undergo the first type of death swell, and then the intracellular organelles[7] disintegrate. The cell membrane then loses its ability to regulate exchanges between an intracellular medium and extracellular medium, so it bursts. The cell literally explodes (the scientific term is lyse), and the intracellular components are released into the external medium. This lysis provokes an inflammatory response that is often accompanied by the formation of a scar.

The situation is completely different in the case of the death that occurs during the weeks after the ligature. Instead of swelling, the cell condenses. The cytoplasm shrinks, and the contents of the nucleus, the chromatin, contracts into a group of separate dark masses. The external membrane then forms invaginations that lead to the formation of cell protuberances. Next comes a fragmentation phase during which these protuberances detach from one another. They contain fragments of cytoplasm, chromatin, and organelles that remain mostly intact. These fragments are finally phagocytosed[8] by the surrounding cells. Unlike the process of accidental, rapid death, the cells eliminated during this slower type of death die inoffensively and silently, without affecting the rest of the organism. There is neither an inflammatory reaction, nor a disorder that disrupts the local metabolic balance.

Kerr coined the term *apoptosis* for that process of cell death by condensation and fragmentation. This term, derived from the Greek, calls to mind the way leaves naturally wither and fall in autumn.[9] Other authors also use the term *programmed cell death*. Hereafter, we distinguish between programmed cell death (or apoptosis), corresponding to a death that plays a physiological role and presents morphological features of conden-

---

[7] Intracellular organelles are a specialized intracellular unit.
[8] This means absorbed and digested.
[9] It turns out that leaves change color and die as a result of cell death by apoptosis.

sation and fragmentation; and necrosis, which corresponds to accidental, rapid cell death due to massive external aggression.

Apoptosis is seen in many organs and in a variety of physiological situations. It plays a fundamental role in regulating organ size—a role that is complementary to the function of cell proliferation (one acting by addition, the other by removal). In the immune system, for example, apoptosis represents an effective mechanism for eliminating auto reactive immune cells, namely those that are likely to produce a destructive immune response against a component of the very organism they are supposed to defend.[10] These cells must be eliminated because they may cause an autoimmune disease. We return to the physiological role of apoptosis in chapter 7.

We have given examples of physiological cell death both during embryogenesis and in adult individuals. These two mechanisms have certain points in common. When we look closely at the morphology of cells that die during embryonic life, we see that, as in adults, many of them die by apoptosis. This is true of all species studied, whether it is insects, birds, or mammals. We even see certain features of apoptosis in bacteria. The preservation of this mechanism across evolution is striking. It suggests that physiological cell death has existed since earliest times, and that the way in which cells die—and even the intricate mechanisms of apoptosis—have mostly been preserved.[11] Hence, we can hope to understand the intimate mechanisms of apoptosis in mammals based on findings obtained from simpler species.

The Mechanisms of Apoptosis: Cell Murder or Cell Suicide?

One element in the mechanism of apoptosis that greatly surprised scientists is that the apoptotic cells seem to actively contribute to their own demise—as if in a suicide—unlike in either necrosis or accidental death,

---

[10] Immunologists call these "self" components.

[11] The definition of apoptosis is based on morphological criteria of condensation/fragmentation. The fact that this morphology has been preserved indicates that the molecular mechanisms have also been preserved, at least in their major lines.

which are more like murder. In necrosis, cells die passively due to an external aggression. However, if we "paralyze" cells by blocking their major metabolic functions, apoptosis can become impossible.[12] In particular, if we inject a cell with a substance that prevents gene expression and protein synthesis, we often[13] observe that apoptosis cannot occur; the cell has to express certain genes for apoptosis to take place. Apoptosis is therefore a phenomenon in which the cell participates actively in its own death, calling to mind the image of cell suicide and the notion of death programmed into the genes of the cell. This comparison does have its limits, however. In certain cases, the cell does not "decide" on its own death, but rather triggers a self-destruction process in response to an external signal—a sort of suicide on command.

As we are going to see, several of these potential "death genes" and "survival genes"—responsible for triggering or blocking apoptosis—have been discovered.

The Nematode: Genes for Programmed Cell Death

The most accurate data that we have on the mechanisms of programmed cell death come from the study of a little worm, the nematode, made up of barely a few thousand cells. From an evolutionary standpoint, the nematode is extraordinarily far removed from humans. However, the process of programmed cell death as it occurs in the nematode can teach us a great deal about what happens in human beings.

Several of the nematode's features[14] make it a valuable organism for studying cell death. First, because it has so few cells (959 somatic cells and 1,000 to 2,000 germ cells)[15] the cells are easier to identify and study individually; and the fewer cells there are to study, the easier it is to analyze cell death. Another major advantage of the nematode is that this roughly one-millimeter-long animal is *transparent*. It can therefore be observed un-

---

[12] Cell metabolism can be blocked only for a short time before becoming toxic for the cell.

[13] Often, but not always.

[14] The species *Caenorhabditis elegans* is the most commonly studied.

[15] See the glossary for definitions of somatic cells and germ cells. In comparison, a human being has thousands of billions of cells.

der a microscope, *alive,* throughout its embryonic development and post-natal life. The division and death of each cell can be observed in real time, from the earliest embryonic stage through adulthood. Building on observations under a microscope, scientists have described an accurate lineage for every cell, a sort of cellular genealogical tree that indicates the mother-daughter relationship for all nematode cells. At the base of this genealogical tree is the egg, which after fertilization and successive divisions will produce the complete animal.

These studies show that 1,090 somatic cells appear during embryonic development of the nematode. At birth, however, there are only 959 cells. This means that exactly 131 cells die during embryonic life via programmed cell death. These cells, when they die, adopt morphology typical of apoptosis. Over the course of about twenty minutes, they become round, condense, and are then absorbed by their neighbors. This programmed death is particularly precise and reproducible, since from one animal to another, the same cell, in the same position, dies at the same stage of development.

From this information then, how can we identify the mechanisms involved? How can we determine which genes play a role in programmed cell death? To answer these questions, the basic idea is to observe mutant nematodes in which the process of programmed cell death has been altered. Such mutants are created by exposing "normal" ("wild-type") nematodes to mutagenic substances that randomly modify the worm's genome, thereby creating modified lines. Certain mutant nematodes manifest excessive programmed death and are generally not viable. There are also viable mutants in which programmed death has been totally suppressed; the 131 cells that should die during embryonic development survive. These types of alterations must involve a mutation of one or more genes participating actively in programmed cell death. These genes can be identified using genetic techniques. The experimental process is complex, but is made easier by the relative simplicity of the nematode genome, which contains "only" eighty million pairs of nucleotides, compared with three billion for humans.

Using these methods, researchers have discovered more than ten genes responsible for the various steps in programmed death. Three of these genes, discovered by the team of R. Horvitz, called *ced 3, ced 4,* and *ced 9* (ced = *ce*ll *d*eath) are involved in the initial phases that trigger programmed death. The genes *ced 3* and *ced 4* are the triggers for programmed death. They provoke the death of the cells in which they are expressed, and mutations that disable them suppress programmed death. The gene *ced 9*, on the other hand, is a gene that inhibits programmed death. In cells where it is expressed, it represses the action of *ced 3* and *ced 4* to prevent the cell from dying. When the action of *ced 9* stops, *ced 3* and *ced 4* can act to kill the cell. Mutations that cause *ced 9* to lose its ability to function result in generalized programmed death that kills the embryo. Mutations that cause an increase in *ced 9* function save cells that should die during embryogenesis.

We now have to ask a key question, the answer to which is one of the most fascinating contributions of this research: Does the discovery of the programmed death genes in nematodes mean that we can identify the programmed death genes in other species, and specifically in humans? Comparing these genes with known human genes provides precise information. For example, *ced 9* has a similar sequence (chemical structure) to a human gene called *bcl-2*. This gene was known to researchers working on cell death before *ced 9* was discovered. In humans, *bcl-2* blocks cell death in a great many circumstances. It probably acts by protecting the mitochondria. Not only does the gene *bcl-2* have a similar chemical structure to *ced 9*, it also has a similar function. A modified line of nematodes was created in which *ced 9* was replaced by *bcl-2*. In these transgenic animals, programmed cell death was identical to that observed in nonmodified animals. So *bcl-2*, a human gene, can replace *ced 9*, a nematode gene. Despite hundreds of millions of years of evolutionary divergence between nematodes and humans, both species have two similar genes that are fundamentally interchangeable. This fact underscores the importance of a process such as programmed cell death, which has retained common features throughout evolution.

The gene *ced 3* is even more striking, since it led to the discovery of a cell death gene in humans. *Ced 3* presents a sequence similar to that of a protease called interleukin-1 beta-converting enzyme (ICE).[16] This protease participates in the metabolism of interleukin-1 beta, a cytokine (chemical messenger). Before the studies on nematodes, ICE was already known, but it was thought to function only as an enzyme involved in the final step

---

[16] An enzyme that cuts and degrades proteins.

of interleukin-1 beta synthesis, which is synthesized in the form of a precursor protein. The sequence of interleukin-1 beta in the protein precursor is inactive because of the presence of extra sequences. ICE cuts these extra sequences, releasing active interleukin-1 beta. The similarities between *ced 3* and ICE suggested that *ced 3* was probably a protease also. This hypothesis is further supported by the fact that one of the mutations in *ced 3* that affects programmed cell death most effectively lies in a region that is almost identical in both *ced 3* and ICE—specifically, the "active site" of ICE, which is most directly responsible for enzyme activity.

The similarity between ICE and *ced 3* tells us that *ced 3* plays a molecular role as a protease. But it calls into question the role of ICE in cell death and apoptosis in humans. Is ICE a cell death gene in humans (and mammals in general)? An initial, partial answer comes from the following experiment: Let mammal cells in culture be modified genetically so that they overexpress ICE. If ICE in mammals has the same role as *ced 3* in nematodes, we can predict that the cells that overexpress ICE will die. And this is exactly what happens.

More generally, it has been demonstrated that ICE is the prototype of a family of proteins, more than a dozen of which have been identified in recent years, that all play a role in some stage of apoptosis. These proteins are all proteases and are grouped under the name *caspases*.[17] Caspases exist latently in the cell in the inactive precursor form (procaspase). When the apoptotic cascade is triggered (via mechanisms that have not yet been fully studied), certain procaspases (described as premature) are converted into their active form. They in turn activate other procaspases, which attack (by digesting) those components that are essential for cell survival, such as proteins involved in DNA repair systems. Procaspases also attack the fundamental structures of the cell skeleton (cytoskeleton), which synthesize cell membranes. This process seriously undermines cell function and structure, and rapidly kills the cell.

The different caspases play different roles depending on the type of cell. To study these roles, researchers use genetic engineering techniques to inactivate certain genes in mice to obtain animals in which certain caspases are absent. As an example, let's return to the mechanisms of programmed

---

[17] Because they have a cystein amino acid in their active site and they cut their targets at the amino acid *asp*artate.

death in the nervous system and brain, so important during embryogenesis, where caspase 3 plays a determining role. In mice that do not express this gene, physiological cell death during embryogenesis is suppressed completely in many regions of the brain. The animal is born with an excessive number of neurons and a hypertrophied brain, and survives only a few days. This effect is limited to the brain. Apoptosis occurs normally in all the other organs, whose physiology is not affected by the absence of caspase 3.

Thus, the studies conducted on the nematode represent a major turning point in our understanding of programmed cell death. They have enabled us to start decoding the mechanisms of this process not only in nematodes but also in humans. They have led to the discovery of a family of genes that code for cell death proteins, the caspases, while also revealing the existence of genes that block programmed death, the *ced 9/bcl-2* family. The chemical and functional similarities between the death genes in nematodes and the corresponding genes in humans demonstrate that the process of cell death is extraordinarily well preserved throughout evolution. A phylogenetic tree (representing the evolution of species) shows us that the branch for nematodes and the branch for vertebrates and humans split more than five hundred million years ago. The fact that nematodes and vertebrates share these programmed cell death mechanisms suggests that these mechanisms appeared before the two branches split. Programmed cell death is, therefore, a very old process in the history of the animal kingdom.

Some mechanisms of apoptosis are even believed to be present in bacteria.[18] This observation is particularly interesting. We know (as mentioned in chapter 5) that mitochondria, which are responsible for producing energy within the cell, are bacterial in origin. They descend from bacteria that are believed to have colonized the ancestors of present-day eukaryotic cells.[19] It is tempting to think that when this colonization occurred, the bacteria transmitted the capacity for apoptosis to the eukaryotic cells. If this is the case, we would expect mitochondria to play a key role in the

[18] X. Wang, "The Expanding Role of Mitochondria in Apoptosis," *Genes and Development* 15 (2001): 2922; B. Mignotte et al., "Mitochondrial Control of Apoptosis: Did Programmed Cell Death Appear after the Endosymbiotic Event Giving Rise to Mitochondria?" (in French), *Médecine Sciences* 14 (January 1998): 54–60.

[19] Eukaryotic cells are cells that have a nucleus, as opposed to "prokaryote" bacteria, which do not.

process of programmed death. Recent experiments confirm this hypothesis and show that deregulated mitochondrial function is involved early on in the process of apoptosis. This deregulation could even be the "point of no return"—beyond which apoptosis is irreversible. Mitochondria are capable of secreting several caspase-activating agents into the cytoplasm in response to external death-inducing signals, or when the cell is subjected to stress or damage that makes it "decide" to do away with itself as quickly as possible. This secretion can be blocked by *bcl-2*, anchored in the mitochondrial membrane, which may be the basis for the anti-apoptotic effect of this factor.

In conclusion, the extraordinary preservation of programmed cell death, written into the living since the dawn of time, underscores its fundamental role. We might say that this phenomenon has a social regulation function within the cells, controlling their number and, when necessary, eliminating individual cells to benefit the organism. We have thus moved from a view in which cell death was the starting point resulting in the death of the individual, leading only to pathology, to a totally different perspective, in which cell death is an integral part of the processes of life. The death of the organism cannot be summed up as the death of its cells any more than the death of the cells can be said to lead necessarily to the death of the organism. On the contrary, cell death plays a role that is often indispensable to the individual. Any failure in this process may lead to serious illness, as we are now going to see.

CHAPTER SEVEN

THE DANGERS OF PROGRAMMED CELL DEATH GONE AWRY

Programmed cell death is indispensable during the embryonic development of organisms. It exists in adult individuals, where it can be activated to regulate the size of cell populations and organs, in the same way as cell proliferation does. But like all physiological processes, apoptosis can malfunction, resulting in either excessive or insufficient cell death. We have recently learned that these failures cause several types of common pathologies, in particular several types of cancers. Apoptosis is also thought to participate in viral infections, neurodegenerative diseases, and immune diseases, including AIDS. This chapter reviews how apoptosis was found to be involved in this wide range of pathologies, and what new therapeutic prospects are available as a consequence.

Apoptosis and Cancer: When Certain Cells Fail to Die

The normal size of an organ at any given moment is the result of a balance between the number of cells that appear by cell division and the number that disappear by apoptosis.[1] At all times, for every organ, we can describe the situation with a simple equation:

[1] Except in accidental death situations.

Number of living cells = number of cells produced by proliferation minus number of cells that die[2]

Either proliferation or apoptosis is triggered by a chemical signal (hormone/growth factor) emitted by the environment. Cancer cells have often lost their sensitivity to these regulatory signals, causing a disruption in the balance and allowing an abnormal increase in the number of cells. This increase may have two different causes, as shown by the equation. It may result from either an uncontrolled increase in the proliferation rate or a pathological blockage in the cell elimination process. Increased proliferation was long believed to be the only reason for tumor formation, and the potential role of a deregulated elimination process was ignored. Cell death was considered to be a marginal process on the physiological level, so deregulations were not believed to have any serious consequences. We now know that this point of view is wrong, and that abnormal suppression of apoptosis is a major cause of cancer formation.

bcl-2: A Cancer Gene, and an Apoptosis Inhibitor

Everything started with the study of the chromosomal changes observed in malignant lymphoma.[3] In 90 percent of malignant follicular lymphomas, there is a chromosomal translocation between chromosomes 14 and 18. In these cells, then, a rearrangement occurs, causing the abnormal juxtaposition of a region of chromosome 18 and a region of chromosome 14. It is not exactly clear why this translocation occurs, but it results in the formation of a new gene. The first part of this hybrid gene comes from chromosome 14, and is composed of an immunoglobulin sequence;[4] the second part comes from chromosome 18 and is composed of the gene *bcl-2*, which was previously unknown.[5] In cells in which the 14–18 translocation has taken place, the gene *bcl-2* is expressed in abnormal quantities. When these results were obtained, nothing was known about the function of *bcl-2*. Initially, it was thought that the overexpression of *bcl-2* caused ab-

---

[2] Specifically by apoptosis.
[3] Lymph node tumor.
[4] A type of antibody.
[5] The abbreviation *bcl-2* stands for *B* cell *l*ymphoma gene 2.

normal cell proliferation, in a pattern similar to what was known for other cancer genes, such as *myc* or *ras*.

To test this hypothesis, experiments were conducted that consisted of artificially increasing the concentration of the protein coded by *bcl-2* in cells in culture.[6] No abnormal proliferation was observed, but an unexpected result was noted: The cells that were overexpressing *bcl-2* were protected from death by apoptosis. This suggests that *bcl-2* plays a role in tumor formation not by stimulating cell proliferation but by *preventing the death of cells that normally should have been eliminated.* To confirm this interpretation, a symmetrical experiment was conducted, consisting of artificially decreasing the concentrations of the protein coded by *bcl-2* in cells in culture. If *bcl-2* is a gene that blocks apoptosis, we would expect a drop in the concentration of *bcl-2* to cause abnormal cell death. This result is exactly what was observed. These experiments were extended to animals. In transgenic mice modified so that *bcl-2* is removed from their genome, several of the organs present a malformation—generally hypoplasia (smaller than normal), the cause of which is massive apoptosis. These mice die young.

All these experiments demonstrate that *bcl-2* physiologically blocks apoptosis. The suppression of *bcl-2* results in abnormally high apoptosis, which leads to the degeneration of the affected organs, as observed in the transgenic animals just described. However, the overexpression of *bcl-2* in an organ pathologically blocks apoptosis; cells that should have died survive, potentially leading to tumor formation, as in the case of malignant follicular lymphoma. It is still not clear how *bcl-2* and other more recently discovered members of the same family function within cells. This protein is concentrated in mitochondria, which are the cell respiration centers. We have seen that *bcl-2* may help regulate mitochondrial function and protect against toxic aggression thereby preserving cell function.

p53: The Cellular Gatekeeper

Tumor-suppressor genes combat the tumorigenic process that can occur within the body. One of the most widely studied tumor-suppressor genes

---

[6] The procedure consists of transferring several copies of *bcl-2* into these cells. Increasing the number of copies of this gene per cell increases the concentration of the corresponding protein, especially because the transferred genes were modified so as to be expressed more efficiently.

is *p53*,[7] which has a direct influence on the process of apoptosis. The *p53* gene was discovered in 1979, but since no one knew exactly what its role was until ten years later, these initial studies did not draw much interest. Subsequently, *p53* mutations were described in tumors from patients with colon cancer. Further experiments showed that when normal cells in culture were caused to express the mutated forms of *p53*, these cells became cancerous. This strongly suggested that the disruption of *p53*'s normal function was involved in the tumorigenic process. It turned out that the role of the nonmutated "wild-type" form of *p53* was to fight against tumor formation. The mutation of *p53* suppressed its antitumor effect, thus facilitating the development of tumors. Two types of experiments led to this conclusion. First, when certain tumor cells in culture or in animal models were made to express wild-type *p53*, tumor progression was halted. A clear antitumor effect of *p53* was observed. Second, when mice were manipulated so that *p53* was eliminated from their genetic code, these animals that lacked the *p53* gene tended to develop cancers more frequently than the wild-type animals. Thus, *p53* was concluded to be involved in suppressing the development of tumors.

We know today that the deregulation of *p53* is one of the primary causes of cancer in humans: fully half of all cancers diagnosed, taken as a whole, are related to mutations of this gene.[8] Mutations of *p53* have been found in more than fifty different types of cancers, including those affecting the breast, brain, lung, colon, skin, and prostate. Seventy percent of colorectal cancers and fifty percent of lung cancers are related to *p53* mutations. In most cases, these mutations are caused by mutagenic agents. Researchers have identified *p53* mutations caused by ultraviolet radiation in skin cells, cigarette smoke toxins in the lungs, and dietary toxins in the liver. These are local mutations that develop as a result of exposure to a mutagenic substance and are present only in the organs where the tumor is located; they are not hereditary. There are certain hereditary forms of cancer, however, such as Li-Fraumeni syndrome, which are due to genetically transmissible mutations of *p53* (which are then present in all cells of the body).

How does *p53* act as a cellular gatekeeper to block tumor formation? The exact mechanism is still the object of intense research, but we do un-

---

[7] So called because the corresponding protein has a molecular weight of 53,000.
[8] This does not mean that a mutation of *p53* alone triggers the tumor process, but that it significantly increases the probability of tumor formation.

derstand the basic concept. DNA damage triggers the cell's alarm system, which can be broken down into three phases. The first phase consists of detecting abnormal changes in the cell's DNA. When this kind of damage is detected, the concentration of $p53$ increases in the cell. The buildup of $p53$ leads to the second phase, which is the immediate cessation of cell division, allowing DNA repair systems to do their work. If the repair is successful, the cell resumes the normal cell cycle. If the DNA repair fails, the third phase of the alarm system is triggered: $p53$ "induces suicide" in the mutant cell by bringing about apoptosis. This process has been demonstrated by causing the overexpression of $p53$ in tumor cells in culture or in vivo. In these experiments, the cells die by apoptosis.[9]

So, $p53$ plays a gatekeeper role against tumors by blocking the proliferation of cells (potentially cancer-causing cells) that have alterations in their genome and by eliminating those that cannot manage to repair the mutation. We now understand why the inactivation of $p53$ can have catastrophic consequences, even when it occurs in only one cell.

We have just discovered a new fundamental role of apoptosis: that of destroying precancerous cells under the control of $p53$. The role of apoptosis in tumor control suggests new therapy approaches in the field of cancer. Anticancer therapies can be based not only on blocking cell proliferation but also on stimulating apoptosis.[10] A simple strategy is to increase the level of $p53$ in tumor cells, which should cause the tumor cell to stop proliferating and die by apoptosis. Several possibilities can be considered for increasing the concentration of cellular $p53$. The first is to find drugs likely to activate expression of the $p53$ gene in the cell. This technique is effective only if $p53$ is not mutated in the tumor, since increasing the concentration of an inactive form of $p53$ would have no therapeutic effect. A second strategy is to transfer a normal $p53$ gene into the tumor cells to obtain large-scale, sustained expression in these cells. This is a "gene therapy" approach, in that the therapeutic agent is a gene and not a standard drug. This strategy is in theory conceivable both for tumors with normal $p53$ and for those with mutated forms of $p53$. However, while from a conceptual standpoint gene therapy principles are simple, putting them into practice requires particularly sophisticated biomedical technology. For ex-

---

[9] In certain types of cells, $p53$ halts proliferation, but does not induce apoptosis.
[10] We now know that most existing anticancer chemotherapy and radiation therapy also act by inducing apoptosis within the tumor.

ample, a gene does not enter a cell naturally, but must do so with the help of a vector. Today, the best vectors are viruses, which have a natural ability to penetrate cells to introduce their genome. In gene therapy, we use "recombinant" viruses that are both disarmed and modified. They are disarmed in that they have been manipulated so that they can still penetrate a cell but have lost their pathogenic ability (i.e., their ability to multiply in the cells they infect). They are modified in that the therapeutic gene to be transferred, in this case *p53*, has been added to their genome. The design and development of recombinant viruses are delicate. Some of the technological problems involved in gene therapy of certain cancers have been resolved. The first human clinical trials involving the transfer of *p53* into malignant tumors are underway and will last several years. These studies will demonstrate whether the use of *p53* as gene therapy against cancer is an effective therapeutic approach.

Apoptosis and Viral Infection

In the presence of a viral infection, a pitiless battle is waged between the infected organism and the aggressor virus. The virus seeks to penetrate the cells of the host organism and feed off of it, the goal being to proliferate and produce as many descendents as possible. In certain cases, this proliferation phase is preceded by a latent period during which the virus sets up camp in the cell before multiplying. During the proliferation phase, the entire metabolism of the infected cell is used for viral multiplication, after which the cell dies of exhaustion. The new viruses then exit the cell, spread, and infect neighboring cells before multiplying and destroying them.

The infected organism, in contrast, seeks to block the infectious process as quickly as possible. Cell suicide is one means of defense against viral infection. The infected cell triggers apoptosis in an attempt to disappear before the virus has had time to reproduce. This strategy limits the spread of infection and protects the rest of the organism. In certain cases, the signal giving the suicide order comes from neighboring cells rather than from the victim cell itself. Immune system agents, the cytotoxic T cells, can recognize infected cells and send them a chemical signal urging them to trigger apoptosis.

Viruses, however, have developed strategies to block apoptosis from being triggered in the host organism; several viruses in fact, have genes that carry out this specific function. The apoptosis program launched by a cell, then, is immediately inactivated by one or more viral genes. We know that in the genome of the Epstein-Barr virus,[11] there is a gene (*BHRF1*) that is similar in structure and function to the antiapoptotic gene *bcl-2*. We also know that this virus can remain latent in the cells it has infected. During this latent phase, it expresses another of its genes, *LMP1*, which stimulates the production of *bcl-2* in the host cell, preventing it from going into apoptosis and thereby protecting the virus that has invaded the cell.

One of the viral inhibitors of apoptosis that has been described widely is the protein p35 of the baculovirus (an insect virus). As described in chapter 6, caspases play a fundamental role in the molecular cascade leading to cell death by apoptosis. The viral inhibitor p35 acts by binding directly to certain caspases and preventing them from doing their job by physically blocking any interaction with their natural targets inside the cell.

Cell Suicide and AIDS

It is also thought that apoptosis participates in the development of AIDS. Before going into detail on the links between apoptosis and AIDS, let's review how the immune system works in general. The immune response involves a large class of blood cell types called lymphocytes, which are produced in the bone marrow. B lymphocytes mature in the bone marrow itself, then migrate to the peripheral lymphoid tissues such as the spleen and the lymph nodes. T lymphocytes migrate from the bone marrow to the thymus where they mature, then to the peripheral lymphoid tissues. T and B lymphocytes play different roles. B cells are responsible for synthesizing antibodies. In mammals there are millions of different antibodies,[12] each of which is capable of recognizing a different antigen[13] for which it is specific. Each antibody, by binding to its antigen, inactivates the antigen either by directly inhibiting its function (for ex-

[11] Epstein-Barr virus is associated with several human cancers, in particular Burkitt's lymphoma.

[12] Each is secreted by a different B cell line (or clone).

[13] An element provoking an immune response, for example a virus, a bacterial toxin, or a component of a microorganism.

ample by physically preventing the antigen from interacting with its target) or by recruiting different destruction systems, such as the phagocyte cells or the "complement" system. T cells do not secrete antibodies but do carry a receptor on their surface (called T-cell receptor) that varies from one T cell population to another. These receptors, like antibodies, are capable of recognizing foreign antigens. T lymphocytes are divided into two classes: cytotoxic T cells, which recognize infected cells via their receptors, then go on to eliminate them; and regulator T cells, which play a fundamental role in the activity of all types of lymphocytes. The latter cells are necessary for the cytotoxic T and the B lymphocytes to provide an appropriate immune response to antigen stimulation.

AIDS is characterized by the massive loss of regulator T (called class 4) lymphocytes, following infection by the HIV virus. This loss paralyzes the immune system, which can no longer respond to foreign antigens. Because of this immune deficiency, the organism becomes vulnerable to infections that a healthy person can easily fight off. The HIV virus has a natural tropism for T4 lymphocytes. In vitro experiments involving T4 lymphocyte cells infected with HIV show the toxic effect of this virus on infected cells. These findings seem to suggest that in AIDS patients, the death of T4 cells is a result of infection by HIV.

It appears that the HIV virus can kill T4 cells even without penetrating them, via a method involving apoptosis.[14] The mechanism proposed is as follows: T4 cells seem to be preprogrammed to go into apoptosis simply by coming into contact with a component of the virus. This contact is enough to "prime" the cells without actual infection or penetration by the virus. The viral component responsible for priming T4 cells for apoptosis may be a protein called *gp120*, which is located on the viral envelope. Next, when these T4 lymphocytes encounter the antigen recognized by their receptors (which generally has nothing to do with HIV), they go into apoptosis. In healthy individuals, on the contrary, an encounter with an antigen signals the T4 cells to proliferate, thereby initiating a response against the antigen.

There is still much debate surrounding both the role of apoptosis in the death of infected or noninfected immune cells in AIDS patients as well as

[14] This suggestion was advanced in 1990 by immunologists A. Capron and J. C. Ameisen. See J. C. Ameisen, "HIV Infection and T-Cell Death," in *Apoptosis and the Immune Response*, ed. C. D. Gregory (New York: Wiley-Liss, 1995), pp. 115–142.

the relationship between apoptosis and the causes of the disease. From a theoretical standpoint, why would a virus induce apoptosis in its host organism cells, especially cells that it has not yet infected? By doing so, it deprives itself of fertile ground for its own proliferation. Why do components of the virus have the ability to trigger the destruction of cells before the virus has been able to make use of them? There are two possible responses. On the one hand, this mechanism may be part of the virus's long-term strategy, in that noninfected immune cells could potentially play an effective role in fighting off the virus. In this case it would be advantageous for HIV to eliminate these cells even without having infected them. On the other hand, apoptosis of noninfected cells could represent an attempt on the part of the cells themselves to mount a defense to limit viral proliferation by eliminating cells that are likely to be used by the virus for this purpose. From the organism's point of view, however, this strategy leads to an impasse, since in the end the individual's immune defenses are destroyed.

From a practical standpoint, if the role of T4 cell apoptosis in AIDS is finally demonstrated, it could open the way for new therapeutic approaches. For example, in addition to an antiviral strategy, there could be an antiapoptotic approach that targets the elements of the apoptotic cascade, in particular the caspases.

Apoptosis and Neurodegenerative Diseases

Neurons, or nerve cells, are particular in that they cannot proliferate. They appear during embryonic life, establishing connections and networks through which they exchange information in the form of electrical and chemical signals. After this phase of embryonic development, a neuron that dies from a toxic aggression, for example, is in most cases not be replaced. The weak regenerative ability of the nervous system contrasts that of most other organs. The skin, for example, can repair itself by rebuilding tissue, leaving behind nothing more than a scar.

The inability of the nervous system to renew itself is the main reason why many brain and spinal cord injuries are irreversible. Neurodegenerative lesions can be caused by an accident, as in the case of cranial or spinal trauma, in which the localized impact kills the neurons in the area involved. The consequences for the individual depend on the region affected. In spinal cord trauma, for example, the location and intensity of the injury determine which limbs are affected as well as the severity of the motor deficit. Massive neuronal death can also occur following a stroke, when there is a sudden interruption in blood flow to certain regions of the brain, causing an oxygen deficit and killing the neurons in the affected regions. Neuron loss can also be caused by progressive pathologies, such as Alzheimer's disease, Parkinson's disease, and Huntington's disease, which are characterized by the progressive death of one or more specific populations of neurons.

The molecular mechanisms of neurodegenerative processes are still only partially understood. In many cases, however, neurological diseases are relatively well described and classified clinically; the cerebral lesions that are associated with the pathologies have been examined in detail in the brains of deceased patients. But the physiopathology—the sequence of biochemical and molecular events leading to neuronal death—of these diseases is still largely unknown. For certain genetically transmitted diseases, such as Huntington's chorea, the mutated gene has been identified, but we do not know exactly how this mutation results in neuronal death. In the case of sporadic conditions such as Parkinson's disease and most cases of Alzheimer's disease, the trigger has not been identified, and attempts to explain how the disease progresses provide only an incomplete picture. Intense research is being conducted in this area, since neurodegenerative diseases are a serious public health problem. Alzheimer's disease is a veritable scourge in developed countries, affecting one out of every ten people over the age of sixty-five, and one out of five over the age of eighty.

One of the causes of neurodegenerative diseases could be the abnormal activation of apoptosis. Neurons do indeed have the ability to go into apoptosis, at least during embryonic development (see chapter 6). Thus, the uncontrolled activation of the cell autodestruction program in the nervous system would obviously cause a neurodegenerative pathology. There are several ways of testing this hypothesis. For example, researchers can examine the brains of affected patients in search of neurons that present morphological characteristics typical of apoptosis. A different approach might attempt to unearth certain mechanisms that are common to apoptosis and neurodegeneration. This latter approach involves studying genes that have a proven role in apoptosis, to learn whether they also play a role in pathological neuronal death.

Let's look at the first approach. Are neurons that degenerate according to an apoptotic process found in the brains of people suffering from this type of disease? Recall that cells that die by apoptosis present a specific morphology characterized by cellular condensation and then fragmentation into apoptotic bodies. It was long thought that the answer was no, but this answer was more often based on theoretical arguments than on experimental observations. In the past, researchers believed that programmed death occurred only during specific physiological processes, particularly those involved in embryogenesis. Moreover, there is a methodological obstacle that makes studying a specific type of cell death difficult in patients. The process of cell death lasts at most a few hours, which means that at any given moment there are only a few cells actually dying; it takes several months to several years for an entire cell population to die. When a patient dies and neurologists can examine the diseased brain, generally all that can be observed is that neurons have died, but not how they died. However, histological techniques have shown the presence of apoptotic cells in the brains of Alzheimer's patients, and similarly in the brains of rodents who died of pathologies similar to human neurodegenerative diseases. These findings are important because they suggest that the abnormal triggering of apoptosis could be the cause of diseases such as Alzheimer's.

Additional arguments support these findings. One of the major features of the brain in Alzheimer's patients is the presence of numerous spherical aggregates measuring roughly one hundred microns in diameter, called "senile plaques." In 1985, the major component of these plaques was discovered: a small, hydrophobic peptide composed of 39 to 42 amino acids and called amyloid peptide (written as Aß). Despite nearly ten years of intense research, the role of senile plaques and Aß is still largely unknown. Likewise, the origin of amyloid peptide and the cascade that leads to its production are also largely unknown. Interestingly enough, it has been demonstrated that Aß can trigger apoptosis of the neurons in the area where it accumulates. This phenomenon could also self-intensify. Several studies suggest that Aß derives from the degradation of a larger precursor and that this degradation seems to occur under the action of caspases. And as we know, caspases are activated as a result of cells going into apoptosis.[15] So it is conceivable that (for an as yet undetermined reason) the brains of Alzheimer's patients' secrete amyloid peptide, which kills neurons by triggering an apoptotic process during which more amyloid peptide is produced, and so forth.

Similar experimental findings link apoptosis and neuronal degeneration for other pathologies. In mice carrying mutations identical to those observed in Huntington's patients, massive neuronal death occurs that has all the characteristics of apoptosis. In this case, cell death can be delayed by caspase inhibitors. Antiapoptotic genes also seem to play a neuroprotective role. In transgenic mice expressing the antiapoptotic gene *bcl-2*, a "cerebral attack" was compared with a similar attack in normal animals. The transgenic mice showed 50 percent fewer damaged neurons than the control animals. Therefore, *bcl-2* is capable of blocking pathological neuronal degeneration, just as it blocks death by apoptosis in other types of cells.

In conclusion, although the possibility that apoptosis plays a role in neurodegenerative diseases is still a working hypothesis, there are several strong arguments that support it. The general pattern could be that at the origin of neurodegenerative pathologies there is an abnormal, aberrant chemical signal, which may vary from one disease to another. In Alzheimer's disease, it could be an increase in the concentration of amyloid peptide. Following a stroke, it could be an increase in the concentration of excitory amino acids. This chemical signal causes certain neurons to go into apoptosis. Different classes of neurons are selectively vulnerable to a particular type of signal such that each signal triggers the death of different neurons and results in a different pathology.

This general model, if accurate, has several therapeutic implications. For example, pathological neuronal death could be prevented by blocking the mortiferous signal, and also by using agents that work further upstream, on the apoptotic process. The advantage of these antiapoptotic agents is that they have a relatively broad spectrum of action and thus can be used to treat several types of pathologies. Potential antiapoptotics include free radical inhibitors and neurotrophic factor. Free radicals, which always seem to be wreaking havoc, are known to play a major role in

[15] See chapter 6.

THE DANGERS OF PROGRAMMED CELL DEATH GONE AWRY **155**

apoptotic processes. Neurotrophic factors, as mentioned in chapter 6, protect against neuronal death (often apoptotic) during embryonic development. Some of these factors have proven to be effective in animal models that involve pathological neuronal death. Clinical trials in humans have been conducted but to date have not provided convincing results. We cannot yet conclude that these factors can be used on a large scale, especially given the numerous methodological problems involved.[16] The coming years, however, should provide a wealth of new information on the possible therapeutic uses of nervous system growth factors.

---

[16] The therapeutic use of proteins like trophic factors poses numerous methodological problems. First of all, these proteins are particularly unstable and degrade quickly after administration. Secondly, the nervous system is protected by the blood-brain barrier that prevents molecules the size of trophic factors from entering the brain via the blood.

CHAPTER EIGHT

DELAYING DEATH

This conclusion relegates the "fountain of youth" to the limbo of scientific impossibilities where other human aspirations, like perpetual motion . . . have already been discarded.

G. C. Williams

The information presented in the preceding chapters seems to account relatively well for aging and natural death throughout the living world. It provides a rich explanatory framework that orients theoretical reflections and which can be tested experimentally. It also suggests practical approaches for studying the basic mechanisms of aging. In this chapter, we examine how, based on this framework and these approaches, we might be able to change the course of aging and prolong life. At this point, the pessimistic view expressed in the citation above may not be totally justified.

Long-Lived Worms

Aging in flies, rudimentary though it may seem, is still a bit complicated to be subjected to a favored approach of geneticists: the search for mutants. Other organisms, however, are excellent candidates, the top two being nematodes and yeast. When times are hard, nematodes stop at a specific phase of their larval development to form a special type of larva, called a "dauer."[1] This larval form is resistant to different types of stress (e.g., heat, ultraviolet radiation, and free radicals) and is capable of surviving for long periods. As soon as circumstances permit, dauer larvae re-

---

[1] German word meaning "duration," pronounced "dower." The main factor that triggers the dauer phase is the ratio between available food and size of larval population.

sume developing and become adult nematodes that seem to be normal in all respects. Their total life duration, however, including the larval phase, has more than tripled. Understandably, some biogerontologists have shown a great deal of interest in these dauer larvae—or more accurately, in the genes that control the choice between normal development and the formation of a dauer superlarva. These genes are called *daf*, for dauer formation. Certain mutations in the *daf-2* gene always orient the larva toward the dauer form, regardless of external conditions. However, this switching error occurs only if the nematode became overheated during its early development.[2] The experimenter simply has to keep a nematode cool to obtain a seemingly normal adult, but one that lives two times longer than its siblings. When all the other worms are dead, 90 percent of *daf-2* mutants are still active.[3] For nematodes that carry an additional mutation in the *daf-12* gene, adult life spans can be three to four times longer.

It is possible, then to extend life span without actually going through the dauer stage. In this specific case (unlike in the case of the development of social insect queens), no external signals are required either. Increased longevity seems to be controlled by a special mechanism that is unrelated to the actual formation of a superlarva and that can act independently from the environment of the nematode; it may be a sort of specialized subroutine inside the dauer program. It is interesting that a single gene can upset the entire life-lengthening mechanism; that appears to contradict the hypothesis that there are a great many genes involved in aging. In fact, *daf-2* is simply a switching device between two predefined patterns of development and aging, both of which depend on many genes.

In research unrelated to the study on dauer larva, the gene *age-1* was identified during the course of systematic research into mutations that increase longevity in nematodes. As adults, *age-1* mutants live roughly twice as long as normal and seem to remain just about as effective from a reproductive standpoint.[4] The *age-1* mutation was the first mutation in the en-

---

[2] This is known as a conditional, thermosensitive mutant. The full effect of the mutation is visible only at a higher temperature (in this case 20 or 25°C compared to 15°C). The benefit of this system is that it allows researchers to study in adults processes that are also involved in early development, without causing too much disruption.

[3] C. Kenyon et al., "A *C. elegans* Mutant That Lives Twice as Long as Wild Type," *Nature* 366 (1993): 461–464.

[4] Contrary to what was initially announced. The strong negative effect on reproduction later turned out to come from a second mutation, in a gene close to but very distinct from *age-1*. However, it could be that a more subtle effect on reproduction is associated with the *age-1* mutation.

tire living world that was found to slow the aging of an organism; the "youth gene" seemed to be within reach. But until 1993, research on *age-1* hardly advanced at all. Then this mutation was found to confer greater resistance to free radicals. Two years later, these same findings were found to hold true for heat and ultraviolet radiation. This coincided with the resistance to stress encountered in flies and also the resistance directly involved in the dauer phenomenon. It was quickly shown that the researchers were on the right track, with the demonstration that *age-1* and *daf* mutations interacted with one another. In fact, the gene *age-1* seems to be a *daf* gene—specifically, *daf-23*, with two types of mutations. The first mutation totally eliminates the function of the gene and always produces dauer larva. The second mutation, initially called *age-1* mutation, only reduces the function of the gene, which prevents activation of the dauer program but not the "increased life span" subroutine.

The gene *age-1* was characterized in 1996.[5] It codes for an enzyme that affects a well-known intracellular signal path, based on the metabolism of phospholipids that form the cell membranes.[6] In fact, the protein AGE-1 was identified because of its resemblance to an enzyme already isolated in mammals. Far from being quirks specific only to nematodes, some of these processes seemed to have equivalents in higher organisms. Several more *daf* genes were characterized in 1997 and were also found to code for proteins that are similar to certain human proteins. One of them, the famous DAF-2 that began our story, acts like an insulin receptor.[7] When activated, it affects sugar metabolism just as the insulin receptor does in vertebrates. When it is blocked, fats accumulate in tissues, similar to the results in dauer larva. In humans, insulin receptor defects are responsible for certain forms of obesity and diabetes. The functional similarities go far, since we find the exact same mutation in *daf-2* in dauer nematodes and in the insulin receptor gene of patients suffering from a hereditary form of diabetes associated with obesity.

Mice models have made it possible to orient and perhaps extend the analogy.[8] Mutant dwarf mice with a growth hormone deficiency live up to

---

[5] J. Z. Morris et al., "A Phosphatidylinositol-3-OH Kinase Family Member Regulating Longevity and Diapause in *Caenorhabditis elegans*," *Nature* 382 (1996): 536–539.

[6] More specifically, they are phosphatidylinositols.

[7] K. D. Kimura et al., "Daf-2, an Insulin Receptor-Like Gene That Regulates Longevity and Diapause in *Caenorhabditis elegans*," *Science* 277 (1997): 942–946.

[8] See, for example, D. Gems and L. Partridge, "Insulin/IGF Signalling and Ageing: Seeing the Bigger Picture," *Curr. Op. Genet. Development* 11 (2001): 287–292, and a critical

50 percent longer than normal. Growth hormone stimulates production in the liver of a factor that resembles insulin, IGF-1 (Insulin-like Growth Factor-1). These dwarf mice have very low blood levels of IGF-1. Conversely, transgenic mice that overproduce growth hormone have high blood levels of IGF-1 and experience accelerated aging.[9] In mammals, IGF-1 seems to be an excellent candidate for controlling longevity in ways that may be similar to what occurs in nematodes. However, direct proof will not be possible until researchers experiment on "cleaner" animals, in which the absence of IGF-1 is not associated with other hormone deficiencies, as it is in dwarf mice.

To go back to nematodes, many questions remain unanswered, starting with the exact nature of the insulin-type signal that acts on the protein DAF-2. The complete sequencing of the nematode genome has revealed no less than thirty-seven genes coding for such signals (compared with half a dozen in most other known genomes). Cells that produce the signal of interest seem to be sensory neurons, most probably taste neurons. The cells that respond to the signal have not yet been identified, but it may be that only a few other neurons are involved. A second signal must, therefore, relay the message to control the age-fighting mechanisms in all the other cells. Once this signal is identified, it may provide new breakthroughs in our understanding of aging as a whole.

What is the evolutionary meaning of the results obtained for *age-1*? Perhaps the most striking aspect is that *the nematode has mechanisms for increasing longevity, but it does not usually use them.* In fact, the mechanisms are actively inhibited except when circumstances require the nematode to take refuge in its dauer form before reaching adulthood. If just one nucleotide of the gene *age-1* is changed, the longevity of the worm is increased. At first glance, this "defect" would seem to be advantageous. Why did it not spread throughout the entire species? The optimization theory immediately comes to mind. Energy invested in the longevity extension process would be diverted from other processes that are more im-

---

analysis by C. S. Carter et al., "A Critical Analysis of the Role of Growth Hormone and IGF-1 in Aging and Lifespan," *Trends in Genetics* 18 (2002): 295–301.

[9] This result could provide food for thought as to the use of growth hormone to fight aging. In this respect, caloric restriction is a much better option, since it kills two birds with one stone: it increases growth hormone levels, while at the same time decreasing IGF-1 levels (doubtless because its action on IGF-1 is located downstream of that of the hormone, which is thus cancelled out). Note, however, that other studies suggest that in humans, IGF-1 may also play a beneficial role during aging.

portant for the reproductive success of the nematode. The identification of DAF-2, which points to a mechanism linked to sugar and energy consumption, also supports this explanation.[10] In normal times, the nematode needs to burn all its sugars in order to reproduce, rather than transforming them into fat stores that have no immediate use. So what if it lives a shorter time! It is also possible that if the AGE-1 protein is not active enough, as in longevity mutants, it does not adequately control the switching mechanism that leads to the dauer larva. These nematodes might then become blocked pointlessly, perhaps irreversibly, in an immature form and be totally sidelined in the reproductive game. The optimal form of the *age-1* gene again seems to result from a compromise that does not ensure maximum longevity but rather reduces the risk of an unwanted switch to the dauer pathway. On the other hand, it is also conceivable that the "normal" form of *age-1* has taken hold by chance, because natural selection didn't care. Indeed, it is only the post-reproductive period that is extended by *age-1* mutations, whereas the fertility period remains unchanged. The normal form of the gene would seem, in this case, to be part of the mutational burden of the species, rather than part of the evolutionary optimization process.

An interesting experiment would be to place the different forms of *age-1* in competition. The initial population would contain, for example, an equal number of nematodes carrying the normal allele and nematodes carrying an allele that confers increased longevity. If increased longevity is accompanied by a reproductive disadvantage, the mutants will disappear from the population over the generations. The difficulty in testing this procedure is reconstituting the environment that prevailed in the wild during the evolution of the nematode—or even just the environment that nematodes encounter in the wild today. Gene performance must be evaluated with respect to these environments.

Experiments on natural selection have been conducted in the laboratory. If the animals have a plentiful food supply, the increased longevity of *age-1* mutants is not accompanied by a disadvantage compared with nor-

---

[10] In a similar vein, a completely different type of longevity gene was discovered (B. Lakowski and S. Hekimi, "Determination of life-span in *Caenorhabditis elegans* by four clock genes," *Science* 272 [1996]: 1010–1013). These genes, called *clock*, seem to control the general metabolism of animals. The mutants age more slowly probably because they live in slow motion. By combining a *clock* mutation and a *daf-2* mutation, we obtain nematodes that live nearly five times longer.

mal nematodes. But if the researchers impose alternating periods of abundance and shortage—a situation that is most certainly more similar to the animals' natural environment—the mutants lose ground.[11] They are quickly eliminated from the population, going from 50 percent to 6 percent of the population after only twelve generations (corresponding to six cycles of abundance/shortage, or a little over two months). Under these conditions, the youngest mutant nematodes lay fewer eggs than their normal siblings of the same age. This inverse relationship between longevity and reproductive performance at young ages certainly supports the hypothesis of antagonistic pleiotropy.

Whatever the case may be, remember that there is no direct advantage conferred by rapid aging. The nematode refrains from delaying its death, even though it could easily do so, not because an earlier death is beneficial in itself but because in the end there is no point in expending too much effort in fighting it off. It is not worth delaying death beyond the point that is truly useful, at least in the organism's usual environment.

All nematode longevity mutants also proved to be more resistant to stress (in the form of heat, ultraviolet radiation, and free radicals). The strong link that reappears here between resistance to stress and longevity is worth closer examination. Recall Kirkwood's arguments in favor of his disposable soma theory. Resources that are allocated to repairing the organism are inevitably less than what is needed to ensure potential immortality. And resistance to stress generally means repairing the damage it causes. The ability to repair seems to be limited by natural selection to a relatively low level, compared to what is biologically conceivable. But what exactly does this mean? The example of the nematode provides one possible response.

A single organism—the nematode—has two preprogrammed levels of resistance, and it normally uses the lower of the two. In normal adults, external conditions may lead to an increase in repair capacities, and by the same token longevity. This brings resistance mechanisms into play and the end result is beneficial, provided that the stress is applied in reasonable doses. The conditions that impel nematodes to take the dauer pathway also bring to mind caloric restriction, which increases longevity in mammals. One of the mechanisms at work could be the induction of

[11] D. W. Walker et al., *Nature* 405 (2000): 296–297.

somewhat more effective repair systems. If so, this supports a general (if theoretical) rule that living organisms have to choose between two different operating modes: One favors reproduction to the detriment of maintenance, while the other does the opposite. The choice depends on the environment encountered. The organism has to hurry to take full advantage of periods of abundance to reproduce, whereas in times of famine or stress it has to economize and protect itself in order to wait for better days to come.

Yet we must remember what we have learned about flies: although it is necessary to tolerate stress well to live a long life, this alone is not sufficient. Some nematode *daf* mutants, although they are more resistant to stress, still do not live any longer. Nonetheless, the link is compelling enough to facilitate the search for longevity mutants. This approach has produced very promising results in the study of aging in yeast.

The Resources of Baker's Yeast

It may seem surprising that we bring up yeast in the context of longevity.[12] It is true that geneticists have shown an interest in this organism for many years. It is the simplest eukaryotic organism and as such is much closer to us than bacteria, which are also single-celled organisms. But it is still a leap of thought to be interested in the age of yeast. It took thirty years for the first observations of this process (in 1959) to be actually put to use. A mother yeast can give birth, by budding, to a limited number of daughter cells. In evolutionary terms, this limit is equivalent to death, even if the old mother does not always physically disappear. Mother and daughter are distinguishable by their size—a difference that is even more visible when the mother is old, because her size continues to increase with the number of daughters she has produced. A few divisions before the mother definitively stops budding, the interval between births increases, making it possible for researchers to predict the moment of death. The age and life span of a yeast can be read by counting the number of bud scars on its surface. However, it is not the accumulation of scars that limits longevity, as was first believed; aging is not simply a matter of mechanical wear. The number of reproductive cycles carried out by a yeast is recorded

---

[12] Even though we discussed yeast in chapter 3 to disprove the belief that aging could not exist in a single-celled organism.

directly on it, and the speed of division provides a good assessment of its physiological age. There is no need to resort to statistics, which can never reveal exactly where a given individual is in the course of its life.

However, this seemingly ideal situation for biogerontologists does have certain disadvantages. The microscopic size of yeast, the impossibility of identifying them individually (especially beyond a certain age), and their low mobility all make it difficult to physically separate individuals or generations. Within one yeast culture, all the ages are mixed up in an exponential pyramid: half are "virgin" daughters, a quarter have already budded once, an eighth twice, and so forth. In practice, the cells are sorted under a microscope to enable researchers to monitor the successive division of each one. After releasing a daughter bud, the mother is isolated from the daughter manually. Otherwise, the mother would quickly be lost in the mass of her own great- great- great-grandchildren!

This somewhat low-tech technique requires a great deal of patience,[13] but it has also allowed researchers to establish the very existence of the phenomenon of aging and its essential characteristics. To understand—at a molecular level—what distinguishes an old mother yeast from a newborn bud required isolation of a large quantity of aging cells. To do this, two U.S. laboratories each developed a different technique.[14] The first is based on the size difference between mothers and daughters. A common technique in biochemistry, ultracentrifuging, is used to separate different sized cells from a mass culture containing hundreds of billions of cells. However, obtaining large quantities of very old cells—namely, those that have divided at least fifteen to twenty times depending on the strain of yeast—is still a painstaking task. To be effective, the separation procedure must be repeated every three or four divisions, before the daughter cells get too big as they themselves divide. As long as a few precautions are taken, each separation phase provides older cells whose age is accurately known.

The second technique is more effective. It is based on labeling the membrane of the cells at a given initial instant. Since the membrane of the buds

[13] To allow the researcher to rest, the yeast are "slowed down" in the refrigerator at night. This doesn't affect their longevity, measured by the number of buddings.

[14] N. P. D'Mello et al., "Cloning and Characterization of LAG1, a Longevity-Assurance Gene in Yeast," *Journal of Biol. Chem.* 269 (1994): 15451–15459; T. Smeal et al., "Loss of Transcriptional Silencing Causes Sterility in Old Mother Cells of *S. cerevisiae*," *Cell* 84 (1996): 633–642.

is formed de novo each time, the daughter cells never inherit the maternal label. The only labeled cells are those that were present initially. After twenty generations, for example, the labeled cells have divided at least twenty times and represent at most one millionth of the population.[15] Their label binds specifically to magnetic microspheres, so a magnet is used to keep them at the bottom of the tube while the other cells are eliminated.

These techniques were first used to identify genes that could play a role in the aging of yeast. Researchers first tested known genes—*ras-1* and *ras-2*. Why those two? Because they help control cell growth and division, and their counterparts in higher organisms are involved in tumor formation. Two essential findings are worth remembering: the activity of *ras* genes decreases with the age of the yeast by a factor of five between the fifth and the twentieth generation; more important, total elimination of the *ras* gene changes yeast longevity by up to 30 percent.

So it wasn't absurd to look for individual genes that would have a considerable impact on yeast longevity, and to focus this search on genes whose activity changes with the age of the yeast. What is the procedure for doing this? Setting aside all technical details, researchers initially play a game of "go fish."[16] A catalog of genetic messages present in old cells is compared with the same messages in young cells. These catalogs initially seem the same, but they differ on a few rare genes. Two genes that are more active in young yeast have been analyzed and named *lag1* and *lag2* (for *l*ongevity *a*ssuring *g*ene). These genes do not resemble any known genes,[17] and their function has yet to be decoded. If you suppress *lag1* or *lag2*, you obtain yeast that appear to be normal but which have life spans that are 50 percent longer or shorter respectively. The two genes "ensure" longevity in opposite directions. Note, however, that the properties of *lag1* disappoint those with naïve expectations. Total inactivation of a gene that is normally less active in the old should reduce longevity even more. Since this is not the case, aging cannot be due to the progressive inactivation of *lag1*. In fact it is probably the other way around. If gene *lag1* reduces longevity, it is probably due to a delayed side effect of its activity in young yeast. Indeed, as Williams stated, the effects of a gene are not limited to its

---

[15] 2^{20} = roughly 10^6.

[16] Also called differential screening.

[17] A gene similar to *lag1* seems to exist in the human species, which would increase the value of the findings obtained on yeast.

periods of activity, since certain effects can manifest themselves much later, when the gene is no longer expressed. More recent results suggest that the overexpression of *lag1* in youth has a negative effect on longevity, whereas restricted overexpression in old age has the opposite effect, as does the total elimination of the gene.

The classic genetic approach has also turned out to be constructive. This approach begins with an examination of the mutations that affect the phenomenon being studied, in an attempt to identify the corresponding genes. But it is difficult to study the longevity of a yeast strain: How can the thousands of mutants that constitute the starting point for any experiment of this type be screened? A fortuitous observation provided a solution. When studying differences in longevity among natural yeast strains, researchers noticed that the longest lived strains were also the most resistant to an extended stay in the refrigerator. The same was observed for resistance to heat shock or famine. This is not really unexpected, since we have seen a link between longevity and resistance to stress several times, for example in fruit flies and nematodes.

After yeast were treated with a mutagenic agent, thirty-nine mutants were found to survive severe food shortage and resume normal growth when food again became plentiful. Eight of these mutants have longevity that is 20 to 50 percent longer than the original wild-type strain. They affect four different genes—*uth1*[18] to *uth4*—and some of the mutations affect the same gene. And surprise, surprise, one of them, *uth2*, was already known by the name of *sir4*! The protein SIR4 has a well-established function: it reduces other genes to silence. Two of its target genes are linked to the sex of the yeast; others are located on the telomeres, the ends of the chromosomes. This information immediately oriented the research. For example, is the *uth2* mutant, alias *sir4*, assigned to the control of sex genes or telomere genes? If either of these, what is the connection to its effect on longevity? This type of link seems to support interesting ideas on the role of telomeres in aging, at least on a cellular level,[19] not to mention speculations that often link death and sexuality.

The first question—regarding the control of sex genes—is easier to answer, and the response is a resounding yes. The control of sex genes and telomere genes is disrupted in the longevity mutant; these genes are no

[18] A homonym for "youth."
[19] See chapter 5.

longer silent. If we reintroduce a normal copy of *sir4* into the mutant yeast, the silence is reestablished. Does a loss of function, because of the mutation explain everything? This is the simplest hypothesis. It is especially attractive because the longevity mutant, by allowing the unrestricted expression of sex genes, prevents sexual reproduction in these individuals (they are called sterile). This is the dream explanation for proponents of a direct link between death and sexuality, since reduced sexuality seems to prolong life. Well, they are in for another disappointment, because total inactivation of *sir4*, which also renders yeast sterile, does not increase longevity, and even tends to decrease it. Conversely, adding a normal copy of *sir4*, although it does reestablish the inhibition of the known target genes—and thus fertility—is not enough to return yeast longevity to normal. The longevity mutation is not simply a loss of function, but it also confers a new function, that may persist even in yeast that become fertile again. Once again, there is no obligatory link between aging and sexual reproduction.

It is not known how this particular mutation of *sir4* increases longevity. Some clues can be found in studying the normal protein SIR4 in old wild-type yeast. In these populations, telomere genes and sex genes are less well repressed—which causes older yeast to be sterile. SIR4 doubtless deserts telomeres and sex genes in favor of the nucleolus, a specialized region of the nucleus. The nucleolus is organized around a region of DNA that contains one hundred to two hundred copies of an essential gene that codes for ribosomal RNA. The nucleolus is one site where RNA is assembled with protein components to form ribosomes, which are crucial elements of cell metabolism. Researchers first believed that the relocation of SIR4 in old yeast could cause them to age, due to the loss of an ability to inhibit (as yet unknown) "gerontogenes" that cause the cell to age. In the context of this model, the longevity mutation appears to reinforce the inhibiting action of SIR4 on these gerontogenes by diverting SIR4 from telomeres and sex genes. In keeping with this model, releasing SIR4 from its telomere functions by artificially shortening the telomeres, far from decreasing yeast longevity, actually increases it. The aging of yeast, then, is not simply a matter of proliferative senescence associated with the progressive shortening of telomeres in cell division. This piece of evidence, therefore somewhat undermines the theory of proliferative senescence as a general basis for aging.

Following these studies, interest in yeast has grown even greater, especially with the discovery of a homologue of the human gene *wrn*, a mutation of which is responsible for Werner's syndrome.[20] This mutation, called *sgs1* in yeast, also leads to a phenomenon that resembles accelerated aging, not only because it reduces longevity but also because it causes premature alterations in the nucleolus, including the relocation of *sir4* to the nucleolus. These alterations seem to be caused by an additional, autonomous copy of the gene that codes for ribosomal RNA; it replicates on its own, exponentially. Fortunately, these internal "parasites" are confined to the mother cell, which preserves the following generations except when they become numerous enough to literally spill over into the daughter cell. They end up representing as much DNA as the entire cell genome. It is easy to imagine how such a structure might make life difficult for the yeast.

In contrast to the initial model, it is now believed that the relocation of SIR4 to the nucleolus is a consequence rather than the cause—of aging.[21] Indeed, by accelerating aging, we also accelerate the relocation of SIR4. The mutated form of SIR4, which slows aging, moves prematurely to the nucleolus: this may be a strategy by the yeast to fight against the multiplication of the parasite DNA. SIR4 seems to play a protective role, with the mutant form being more effective because it reaches the nucleolus earlier. The protein SGS1 seems to intervene upstream by delaying the formation of the first copy of parasite DNA. In support of this idea, SGS1 is concentrated in the nucleolus (like its human homologue WRN) and is capable of preventing rearrangements in the DNA; this evidence fits in with the supposed function of SGS1. Do chromosomal instabilities, with the invasion of parasite DNA from the genome, play a major role in human aging? This remains to be seen. The fact that this question arises based on studies of yeast, a distantly related single-celled organism, again underscores how certain fundamental mechanisms of the living have been preserved throughout hundreds of millions of years of evolution.

More recent findings have added an even more unexpected dimension, if this is possible, to the yeast model. One of the *sir* genes, *sir2*, has homo-

[20] Werner's syndrome mimics certain aspects of aging, in an accelerated form (see chapter 5).
[21] Another example of the difficulty of distinguishing the cause from the effect in biology . . .

logues in practically all living species, from bacteria to mammals. The protein SIR2 has turned out to be an enzyme specialized in the modification of chromosomes.[22] In fact, it inactivates them. The genes located in the modified regions are reduced to silence. SIR2 can thus control the global expression of the genome. Up to this point, however, nothing suggests that it plays a universal role in aging. What tipped the researchers off was that SIR2 activity seemed to depend on the metabolic condition of cells. A slow metabolism, which increases the longevity of yeast and many other organisms, stimulates SIR2. And the overexpression of SIR2 also increases longevity. From there to imagining SIR2 as the link between metabolism and longevity was only a small step, and it was quickly taken: in the absence of SIR2, caloric restriction no longer increases longevity.[23] The next step was to move from unicellular to multicellular organisms, thanks to transgenic nematodes having several copies of their own *sir2* gene. Their longevity is as much as 50 percent longer than normal. What's more, this increase involves the *daf-2/age-1* pathway discovered previously.

Until we get confirmation that this genetic mechanism is a universal aspect of animal aging in general, and therefore probably human aging as well,[24] these studies on yeast refine the essential condition that Weismann stipulated for the appearance of "natural" death—namely, the separation between germ and soma. In yeast, this condition is reduced to its most simple expression: the asymmetry between the large mother cell and its daughter bud. The minimal differentiation makes the loss of individual immortality acceptable. The "germ line," created by a series of first buds, is potentially eternal, even if each individual yeast is incapable of budding more than twenty or thirty times. In fact, a mother yeast would need to divide only twice in order for the population to increase exponentially, with each daughter in turn producing two buds, and so forth. The demographic potential is that much greater because each mother can engender more daughters. However, the contribution of the mothers quickly be-

[22] More specifically in the deacetylation of histones, proteins that are DNA's main partners in forming chromosomes.

[23] The absence of SIR2 reduces the longevity of yeast for the same reasons that the absence of SIR4 does. The strain used here involves another mutation, designed to compensate for this reduced longevity, but which alone does not alter the response to caloric restriction.

[24] Discussed by L. Guarente, "SIR2 and Aging: The Exception That Proves the Rule," *Trends in Genetics* 17 (2001): 391–392.

comes negligible compared with that of their exponential progeny. What would be the point of prolonging this contribution indefinitely? As Weismann put it, this would be an "unnecessary luxury." Conversely, what is the "necessity" that determines the actual limit, which varies from one strain to another? Why do some strains prevent the formation or accumulation of parasite DNA more effectively than others? This question goes back to the current evolutionary debate between mutation accumulation and optimization. The identification of longevity genes, and especially their role at a cellular level, may provide new answers. As in the case of the nematode, it would be interesting to place the wild-type strain and the long-lived mutants in competition to see whether by living longer, the reproductive success of yeast suffers in the long term.

"Gene Therapy" against Aging?

It seems, then, that overexpression or modification of a single gene can increase longevity in yeast. Would this type of approach work in a more complex, multicellular organism, which doubtless faces many more causes of aging? It is logical to look at oxygenated free radicals in this context. We have come across this promising line of thought several times before. Free radicals are probably involved in the effects of caloric restriction, which for now provides the surest means of prolonging animal life—from protozoa to primates, and most likely in humans as well. We can test this idea, by "building" organisms that have a greater ability to eliminate free radicals, in the hopes that they will have an increased life span. This is what Orr and Sohal observed in 1994 on fruit flies to which they had added an additional copy of genes coding for two key enzymes in free radical detoxification, catalase and superoxide dismutase (SOD). These transgenic flies fulfilled all the hopes of their "parents," in that they lived up to one third longer than normal. In this experiment, fifteen separate lines were created. Longevity increased in eight lines, remained unchanged in six, and decreased in one. Of the four control lines, one had very low viability, whereas the longevity of the others was normal.

Statistically, at least, the effect of the transgenes studied is clear. The research verified that the activity of the corresponding enzymes, SOD and catalase, was higher (by 30 and 50 percent respectively) in the long-lived transgenic lines than in the control lines. Aging, defined by the increase of mortality rate with age, also decreased by the same amount in the long-lived lines. These lines accumulate oxidative damage in DNA and proteins less rapidly, even though their metabolism is more active. These findings support speculation on interesting therapeutic possibilities, which we discuss at the end of this chapter. It is important, however, to bear in mind the evolutionary paradigm when evaluating how such possibilities may actually be put into practice. Why, for example, are flies not normally better able to detoxify, which would enable them to live longer? All they would have to do is increase the activity of two little genes out of more than ten thousand. This solution, however, was not selected in the course of evolution. As in the case of longevity mutations in the nematode, the ideas of Williams and Kirkwood provide an explanation: Perhaps, when all is said and done, increased detoxification would not be such a great advantage. Experiments are underway to find possible harmful effects of detoxification, specifically on fecundity.

In the area of aging, other transgenic flies provide a good illustration of the dangers of an excessively simplistic approach. The story starts in 1989. A highly reputed Swiss laboratory published an article on a line of fruit flies whose life span had been increased by nearly half. The gene that had been added to them, *EF1-alpha*, plays a crucial role in the synthesis of all cell proteins. The result was thoroughly plausible: In flies, the activity of *EF1-alpha* normally decreases with age; and more general theories on aging cite progressive problems in protein synthesis. The article made headlines in the mainstream press—for example 'Grim Reaper' Foiled by Science."[25] Later, the comments were toned down a bit: "Unfortunately, the Swiss researchers were not able to reproduce their first findings. Subsequent experiments were disappointing."[26] When it was finally measured, the expression of *EF1-alpha* due to the transgene proved to be insignificant. Most of the other *EF1-alpha* transgenic lines constructed afterward did not manifest increased longevity under the same conditions. It now seems clear that no conclusions can be drawn from the 1989 study.

This story illustrates the practical importance of the theoretical evolutionary framework. One of the articles that challenge the first interpreta-

[25] *Le Figaro*, 6 February 1990.
[26] *Le Nouvel Observateur*, 12 January 1995.

tions involving EF1-alpha stated that "each person's approach is biased by his education and personal history [as for us] we would not have looked for a single gene that has a major effect, because our vision was modeled by the evolutionary theory of aging, which suggests that normal aging is determined by numerous genes with small individual effects . . ." Thus each discipline was holding one end of the solution, and only by working together could they succeed.[27] An evolutionary perspective suggests repeated verification before claiming that overexpression of a single gene increases longevity, as in nematodes or yeast. A detailed analysis of Orr and Sohal's findings reveals certain complexities. When they introduced only one of the free radical detoxification genes, either SOD or catalase, they often obtained flies that lived a shorter time than normal. Even with these two genes, the effect is not always guaranteed, since it shows up in hardly more than half the lines tested. The effect is nonetheless very striking, bolstering both the desire to understand the primary mechanisms of aging and the hope of combating them.

Fooling Our Genes?

Richard Dawkins gave the neo-Darwinian vision of evolution a deliberately provocative form, that of the "selfish gene." The use of such emotionally loaded words causes a bit of a problem, as Dawkins himself admitted. Dawkins was careful to say that the "selfishness" of genes does not imply either conscience or intention. What is interesting for us is that he underscored an important issue, also mentioned by Williams: the environment of a gene is not simply the external factors surrounding the individual carrier of the gene. It also includes all the other individual's genes, and the rest of its "internal" environment, which Williams called the "somatic environment." This can in part be considered the result of interactions between the genes and the external environment.

Such a view has interesting conceptual implications for those of us who wish to study the causes of aging. We have already seen that a single gene is not likely to be "the answer." Even in the case of the transgenic SOD and

[27] S. C. Stearns et al., "Effects on Fitness Components of Enhanced Expression of Elongation Factor EF-1-alpha in *Drosophila melanogaster:* I. The Contrasting Approaches of Molecular and Population Biologists," *American Naturalist* 142 (1993): 961–993.

catalase fruit flies—to date, the most convincing argument for gene therapy against aging—the results are hit or miss. Moreover, the researchers did not in all cases look for the potential side effects of increased longevity, as predicted by the theories based on Williams's ideas. It is very possible that these longer-lived flies may be less healthy or less able to reproduce than normal flies. We must remember that natural selection fashioned the entire genome, giving precedence to the development of the corresponding soma up to the moment of sexual maturity—and without regard for the fate of the soma after reproduction. So we probably shouldn't expect too much from changing a single gene.

Dawkins made an original suggestion, and in subsequent editions of his book he expressed his surprise that it was not taken more seriously. He proposed "fooling the genes by making them believe that the body they are in is younger than it really is." To put it another way, since natural selection has retained genes for their performance in the young, let's try to simulate the "internal chemical environment" of a young body to provide these genes with the environment to which they are best adapted. For this challenge to have a reasonable chance of success, however, only a small fraction of the internal environment can count, because renovating the entire system would boil down to reversing aging, and the problem would already be solved. What Dawkins seemed to have in mind was the existence of one or more special biochemical signals—a hormone concentration, for example—that would tell the genes the age of the organism and thus determine the moment when their late-onset harmful effects would appear. Depending on the initial value of the signal, and how fast it changed, aging would affect each individual at a different rate, notably based on its interactions with the external environment. This concept corresponds to the idea of an aging biomarker.[28] We have seen that for a biomarker to be useful in monitoring individual aging, it need not have a direct role in age-related physiopathological changes. Likewise, in the hypothesis we are currently discussing, the signal that acts "in spite of itself" as an indicator to genes can have its own function that is unaffected by aging and has no direct effect on aging.

These conditions are not purely a matter of form, first of all because a biomarker for human aging may have been identified thanks to the chem-

[28] See chapter 2.

ical DHEA,[29] *a precursor* for sex hormones. More specifically, the bio-marker is the change in DHEA concentrations with age. For any given subject, blood levels of DHEA decrease continuously with age. Moreover, it seems that on average, the individuals whose levels drop most quickly are more likely to develop certain age-related diseases. Experiments on animals and clinical trials suggest that the administration of DHEA, whose biological function is unknown, could improve the general health of old subjects.

However, it is much too early to talk about the fountain of youth, or even to use the only slightly less enthusiastic expression "the dream drug for aging well."[30] What we want to underscore is that speculations such as those of Dawkins, which seem very theoretical, enable us to keep an open mind when looking at complex experimental data. Indeed, we must acknowledge the possibility that DHEA does not have any direct action on a specific aspect of aging. It may simply be an environmental signal that affects the genome as a whole because it was selected for its performance in a young organism—that is, one with a relatively high concentration of DHEA. This outlook has the advantage of considering the genome as a whole. It explicitly stresses the fact that all genes are in the same general bath,[31] and further, that the composition of this bath depends greatly on how the entire genome works. Take the *ACE* gene, for example. One of its alleles, which is associated with a decreased risk of heart attack, seems beneficial in young individuals. However, carriers of the other allele are overrepresented among centenarians. This second allele could confer an advantage in a very old organism. Interactions between genes, and their dynamics as an individual increases in age, must certainly be taken into account when studying aging.

And here we come up against the same old paradox: why in the world didn't natural selection program into the genome the ability to maintain a young somatic environment, such as one with higher concentrations of DHEA? The neo-Darwinian framework can help us resolve this paradox in two different ways. If we follow the path proposed by Medawar, the

[29] Abbreviation for dehydroepiandrosterone, discovered by E.-E. Beaulieu more than 30 years ago and much hyped in the media in 1994–1995.
[30] *Le Point*, 7 January 1995.
[31] In all strictness, the somatic environment can differ from one type of cell to another within the same organism, if only because of the specialized biochemical reactions that take place in the different tissues.

decrease in DHEA levels in old individuals could result from late-onset mutations that accumulate in the genome of the species. Williams would be more likely to look for the advantage provided to young individuals of a genetic configuration that causes DHEA levels to drop with age. In both cases, the use of this signal could simply be chance and opportunism on the part of natural selection. For example, there would need to be only a minimal correlation between DHEA levels and the reproductive probability of an individual. Any pairing between the genome and DHEA levels would then be advantageous, because the organism could more easily adjust its functions to its future prospects—specifically, those regarding the compromise between maintenance and reproduction. But be careful not to deduce from this that there is a deliberate aging program resulting from specific and coordinated activity. Such a program would be incompatible with the evolutionary theories of aging, which state that natural selection simply adjusts the speed at which the late-onset harmful effects of genes appear in order to optimize the reproductive abilities of the organism depending on its environment. An internal orchestra conductor that sets the tempo for aging would be useless.

Deciding which of the two major evolutionary theories of natural death is right, whether it be mutational accumulation or antagonistic pleiotropy, is not of academic interest only. To fully understand this, let's look at the different predictions these theories make on the evolution of human longevity. The demographics of our species have changed a great deal over the last two centuries. Mortality rates have dropped everywhere—in some cases spectacularly—and not only at the youngest ages. As in the case of Austad's island opossums, natural selection can now operate on the late-onset effects of genes, effects that until now rarely had the chance to manifest themselves. The spread of natural selection's action to older ages is also favored by the sociocultural trend to delay having children. The priority is no longer to have as many children as possible, as early as possible. This sociocultural evolution probably reinforces biological evolution by anticipating some of the trends that can be predicted based on Austad's observations of opossums, or on Reznick's observations of guppies transplanted into more protected environments.[32] From an evolutionary standpoint, modification of the reproductive timetable and drop in ex-

[32] See chapter 3.

trinsic mortality are two factors that should tend to select for better and better soma maintenance mechanisms over the generations. But, depending on whether we go along with Medawar's or Williams's vision, the consequences are very different.

According to Medawar, natural selection simply has to push back the age at which those genes that contribute to the natural death of our species produce harmful effects. Remember that this does not necessarily mean mutations in these genes themselves. In theory, natural selection can also operate on genes that control the activity of others. In contrast, according to Williams and proponents of the life history optimization theory, suddenly tipping the scales toward preservation and especially reproduction in older organisms is likely to be accompanied by negative effects in the young, at least in the first generations. Human aging being the byproduct of a long selection process in favor of the highest possible fecundity at sexual maturity, any inversion of the process can negatively affect fecundity (specifically in very young adults) and perhaps even development in childhood.

In support of this prediction, recall the works of M. Rose and others who studied the evolution of longevity in fruit fly populations.[33] They suggested that the progressive increase in longevity from generation to generation—a result of later procreation—occurs to the detriment of the performance of flies at the start of their adult life. Could this be the primary cause for the drop in human sperm quality that has allegedly been observed in recent decades?[34] No, because the drop is much too rapid to be caused by evolution, which takes a minimum of ten generations to have an effect. Moreover, we must acknowledge that the human species has essentially removed itself from the effects of natural selection. We have almost no predators other than ourselves, and medicine has thankfully reduced the impact of genetic differences among individuals, specifically regarding their ability to have children. This effect could, in the long term, lead to a general drop in the reproductive performance of our species, because it would no longer be controlled by selection. The recent upheavals in our life cycle could then counter this drift. Overall, these changes amount to raising the selection bar by requiring individuals to

[33] See chapters 4 and 5.
[34] We say "allegedly" because this finding is regularly challenged. The research methodology is very tricky, and it is difficult to eliminate all bias.

reach an older age before procreating—which may in turn create a demand for even better medical technology. The end result of the mutual interaction between life cycle and medical progress is difficult to foresee. In any case, as L. Partridge and N. Burton wrote, "unfortunately, the contemporary authors and readers of this article have almost no chance of seeing the results of this experiment."[35] Hence, we have resorted to the nematode and the fruit fly.

Do We Age More Slowly As We Age?

This surprising question is actually the topic of heated debate. Since the 1990s, several series of findings have cast doubt on the validity of the Gompertz law (described in chapter 2) in the oldest ages in flies, nematodes, and humans. The curve that describes the increase of mortality rate with age—the parameter that defines the speed of aging—tends to flatten out. Instead of obeying an inexorably exponential law, the increase in mortality rate seems to slow gradually. But keep in mind that this does not mean that mortality begins to decrease, just that it increases less quickly or stabilizes. Moreover, this slowdown seems to occur at a very old age—around eighty-five to one hundred years in humans, according to the authors. The majority of the population has disappeared long before being able to take advantage of this relative reprieve.

It is not easy to demonstrate this phenomenon and even harder to quantify it accurately. To do so requires studying populations large enough for the number of individuals in these later ages to remain significant. Most researchers have accepted these findings, but their significance is far from being unanimously accepted, except on one point: Since the Gompertz law has no known theoretical justification, neither does its "violation." In fact, a violation of the Gompertz law should be less surprising than its degree of validity. Rose's team attempted to explain both the validity and the violation of the law, using mutation-selection models.[36] These models include one of the essential ingredients of evolutionary theories of aging: mutations that have different effects depending on the age of the organ-

---

[35] "Optimality, Mutation and the Evolution of Ageing," *Nature* 362 (1993): 305–311.
[36] L. Mueller and M. Rose, "Evolutionary Theory Predicts Late-Life Mortality Plateaus," *Proc. Nat. Acad Sci. USA* 93 (1996): 15249–15253.

ism. In all cases, after a few hundred simulated generations, the mutations lead to mortality rates that grow according to the Gompertz law initially but then hit a plateau. In other words, the mortality rate becomes constant after a certain age that is reached by less than one percent of individuals. Other theoreticians have used a more physiological approach, attempting to relate reproduction and mortality in the context of the disposable soma theory.[37]

Depending on the parameters included in these models, optimization by natural selection can lead to very different mortality curves. The curve that corresponds to the Gompertz law is only one particular case, unlike what is shown by Rose's simulation results. Perhaps these curves are based on hypotheses that are too restrictive. But the trend toward constant mortality in these oldest ages also appears in these curves. Both groups of researchers explain this phenomena in roughly the same way: there is an age after which reproductive value is so low that natural selection cannot tell the difference. All the oldest ages of these organisms are the same in terms of the late-onset harmful effects of the genes present. These genes do not accumulate preferentially at the top of this age group, which is already so extremely old.

Other researchers, notably demographers, defend the Gompertz law. They stress that a mortality curve is by nature statistical. It summarizes the future of an entire population, but none of its members will necessarily conform very closely to the curve. A flattening of the curve does not mean that the rate of aging of each individual changes with age. The effect seen on old ages could just as well be explained as the early elimination of the majority—the most fragile individuals who age more quickly—and by the preferential survival of a more hardy minority who age more slowly. Each of the two subpopulations respects a Gompertz law, with its own parameters. Aging seems to slow down when the fragile majority has almost disappeared and the population is composed primarily of the hardy survivors. In theory, this explanation is plausible because natural populations are generally very heterogeneous. These experimental results can be reproduced with a mathematical model that superimposes several subpopulations. The problem is that the number of populations is a bit arbi-

[37] P. Abrams and D. Ludwig, "Optimality Theory, Gompertz' Law, and the Disposable Soma Theory of Senescence," *Evolution* 49 (1995): 1055–1066.

trary and above all, very different Gompertz parameters must be applied to each population. As a result, differences in aging rates between the sub-populations become comparable to the differences observed between mice and men, and such disparity within one species is hardly plausible.

Does the fact that mortality stops increasing at the oldest ages contradict the notion that an absolute limit to longevity exists within each species? It does, in the opinion of demographer James Vaupel, who suggests—optimistically—that more and more of us may follow the example of Jeanne Calment and live past the barrier of 120 years, long considered the final frontier. He may be right, especially because the latest generations will have had much better living conditions from birth.[38] But there will still be very few who live past 120 years, since the mortality of centenarians, although it does not increase at that age, is still very high (around 25 percent per year). The traditional notion of absolute limit to longevity seems to be more questionable, in any case. This debate is worth following closely in the coming years, because it could challenge our concept of aging—and there is certainly no harm in dreaming!

Living Longer by Eating Less?

These long-term speculations aside, perhaps we already have a relatively simply practical answer for prolonging life. Osborne in 1917, and McCay in the 1930s demonstrated that by putting laboratory rats on a strict diet, it was possible to increase their life span by 10 to 30% percent compared with those fed at will. These findings were long considered to be a curiosity of the laboratory environment, before being reexamined in the 1970s. Walford and Weindruch reported in 1988 that a caloric restriction of 60 percent increases the maximum life span of rats by more than 50 percent. A similar phenomenon was observed in many different species, including fish, spiders, mosquitoes, and flies, suggesting a general link between calorie intake and longevity. It is remarkable that animals (rodents in particular) fed a low-calorie diet also present fewer serious diseases

[38] It is a striking coincidence that the number of centenarians began to increase one hundred years after actual wages began to rise. The correlation with the decrease in infectious diseases or infant mortality is much weaker. (M. Perutz, "Long Live the Queen's Subjects," *Philos. Trans. R. Soc. Lond. B* 352 [1997]: 1919–1920.)

than their nonrestricted fellow creatures. The incidence of cancers, cardio-vascular diseases, and kidney diseases is much lower. Moreover, most of the physiological, hormonal, biochemical, and behavioral signs associated with aging are also delayed. The rats on the diet have better cognitive abilities and their immune systems are in better condition.

Why does caloric restriction increase life span? First, we should make a distinction between the hypocaloric diets tested in these experiments and undereating. The controlled low-calorie diets used in these experiments contain all the nutritional elements essential for life (i.e., protein, vitamins) but in a low-calorie form—hence the term caloric restriction—whereas underfed individuals lack these essential factors, which obviously leads to a dramatic decrease in life expectancy. One of the first hypotheses to account for the effects of caloric restriction was that it caused a general slowing of growth and maturation, thereby delaying the onset of aging. This hypothesis has since been abandoned, at least as concerns mammals, since we know that longevity can be increased even if caloric restriction is initiated after the animal has reached adulthood. A more satisfactory explanation is based on the theory linking aging to free radicals.[39] Because free radicals are dangerous by-products of energy consumption within the cells, any decrease in calorie intake should automatically result in a decreased production of these toxins, and consequently slowed aging. In keeping with this hypothesis, the concentrations of free radicals are lower in rats subjected to caloric restriction, especially in organs such as the muscles, heart, and brain, which are particularly exposed to oxidative damage (since they are composed of cells that do not divide and, moreover, consume large quantities of oxygen).

In recent years, the large-scale genome sequencing of numerous species has provided new arguments indicating that caloric restriction slows aging. It is now possible to simultaneously measure the expression of thousands of genes in an organism, or even in just one organ. The "gene expression profile" obtained reflects the activity of the genome under the particular conditions prevailing in the tissues in question. Experiments have been conducted in organisms of different ages, some of which were subjected to caloric restriction. The changes in their profiles with age

[39] See chapter 5.

show that the activity of most genes is practically unrelated to age. Aging, then, does not seem to result from a massive deregulation of the genome. Among the minority of genes whose activity does change, researchers were able to identify a subset for which the changes are slowed, or even eliminated altogether, in calorie-restricted organisms. In fruit flies, it was even possible to establish a correlation between this slowdown and slowed aging overall (measured as a slower increase of mortality with age).[40] In other words, the activity of these genes seems to reflect the state of aging, and thus the physiological age—rather than the chronological age—of the organism.

These data, which are just beginning to appear, offer a valuable guide for objectively evaluating antiaging strategies by comparing gene expression profiles with and without treatment. Although it is still too early to draw specific conclusions on the mechanisms at work, some constants have appeared that largely confirm the current models. Among the genes identified, many are involved in responses to stress (especially oxidative stress) and in cell proliferation. In the liver of aging mice, the expression of the apolipoprotein E gene decreases.[41] This may be related to an increased incidence of atherosclerosis. Remember also that human genetic studies (see chapter 5) showed that this gene could be a potential determining factor for longevity.

The gene expression profiles of different organs change differently with age, according to the various stresses that the tissues face, depending on cell renewal rates and depending on energy consumption. Based on these profiles, caloric restriction seems more effective for some organs than for others. But perhaps the most telling result for us is that it is not necessary to "suffer" caloric restriction for an entire life to obtain some of the benefits—in mouse livers, at least. Mice put on a diet lasting only four weeks show roughly two thirds of the changes seen in mice subjected to caloric restriction for their entire lives. Although these beneficial changes do not necessary cancel out all the deleterious effects that may have occurred previously, the findings do suggest at a minimum that the "slowed aging" mode can be instituted at any age.

[40] S. D. Pletcher et al., "Genome-Wide Transcript Profiles in Aging and Calorically Restricted *Drosophila melanogaster*," *Current Biol.* 12 (2002): 712–723.
[41] S. X. Cao et al., "Genomic Profiling of Short- and Long-Term Caloric Restriction Effects in the Liver of Aging Mice," *Proc. Nat. Acad. Sci. USA* 98 (2001): 10630–10635.

What lessons can we draw for the species *Homo sapiens?* First of all, remember that the results just described were obtained in the laboratory—that is, in a controlled, confined, and protected environment, where the environmental parameters are identical for each animal. Moreover, all the animals used in each experiment are usually genetically identical. This state of affairs is extremely different from the "natural" human situation, in which other highly variable factors (e.g., physical activity, circadian rhythms, sexual activity, psychological stress, and genetic differences), are likely to modulate the role that caloric intake plays in longevity. It will take several years before the caloric restriction studies begun on primates provide conclusive results, although the preliminary findings are encouraging. There are no decisive studies in humans, however, but there are some interesting observations. For example, the Japanese island of Okinawa has forty times more centenarians than the rest of Japan, and it just so happens that the diet of the inhabitants is low in calories while still being very nutritious. This example, of course, is not a definitive demonstration of the effects of caloric restriction on longevity. Other factors including the lifestyle and genetic structure of the population in question may also play an important role.

Whatever the case may be, we can all test the effectiveness of caloric restriction on ourselves, provided we take the essential precautions to avoid any nutritional deficiencies. If we adhere strictly to such a diet, we may live longer. But everyone will have to decide for him or herself whether the permanent discomfort and monotony of a very low-calorie diet simply makes life not worth living!

CONCLUSION
Good Reasons to Die?

Any trace of *positive value judgment* is a bad sign for a mind whose goal is to be objective.

G. Bachelard

Throughout this book we have tried to describe the place that death occupies in life sciences today, on a cellular level and for an entire organism. The primary mechanisms at work have shown themselves to be more accessible to experimentation than we feared. And even though these mechanisms are not completely clear, we believe that the results obtained over the past twenty years are fascinating enough to be worth presenting outside the specialized scientific community. Moreover, they fit into conceptual frameworks that enable us to outline the medium to long-term prospects for the future—not only for research but also for human longevity—thus providing food for thought for each and every one of us.

We would like, however, to underscore the complexity of the subject. There are still many questions for which we have no definitive answers—life cycle optimization versus mutation accumulation, the existence of nonaging species (whose members are potentially immortal), slowed aging at the oldest ages, an individual biomarker for aging, and possible links between cell senescence and aging. All of these are controversial issues, demonstrating how fragile certainty can be in this area.

One fact seems clear: the existence of aging and natural death does not result from any particular advantage that it may have for an individual or for a species. At the risk of sounding simplistic, we suggest that natural death has no value in and of itself; its existence is simply the result of a central biological pointlessness to repair systems that would prevent aging. All living organisms—even nonaging ones, even "amortal" ones—are doomed to exist temporarily, because they are perishable. An accident, a disease, a predator, can interrupt the course of their lives, and the time they have to procreate cannot be extended indefinitely. Natural selection "judges" each organism by the yardstick of procreation (which includes the care and upbringing provided to its descendants). The goal is successful reproduction, even if it leads to a progressive shortening of a species' life span over the course of evolution. Rushed by the fatal end (which was originally due only to external causes), the organism simply does not have the time or resources to make perfect repairs. Without going to extreme examples of mayflies or salmon, this arrangement leads the organism to neglect itself just enough so that aging and natural death occur. Faced with the mortal threat that will eventual kill it, it has to deal with the most pressing matter and successfully complete its task. Death sets the true priorities of the organism, and natural selection ensures that they are respected. In general—we hope that we have demonstrated this in our book—amortality cannot be one of these priorities.

Based on this observation, it is interesting to analyze the conceptual obstacles that "thanatobiology" has encountered. We believe that these obstacles are related to the fact that since the last century, science has been expected to "make sense of things." In our opinion, the very idea of basing a "message" with a social dimension on scientific works should produce a healthy mistrust. This opinion has also been expressed recently by others, including F. Jacob, "[The theory of evolution] explains what we are, but it doesn't say what we should do and why we should do it!"[1] and P.-H. Gouyon: "I have morality on the one hand and a biological explanation on the other. And I have no lessons to learn from nature on what I should do."[2] If there is in fact any message, we must be careful to formulate it in the most restrained and prudent way, even at the risk of disappointing an audience looking for hope, comfort, or simply consolation.

Epistemological Obstacles

In this book, we have sketched out the hesitations and the misconceptions that have marked the history of scientific ideas on death. Why has this path been so difficult? First of all, we must acknowledge that it takes

[1] *La Recherche,* March 1997, p. 24.
[2] *La Recherche,* November 1996, p. 93.

an immense effort to understand death scientifically, since an intimate awareness of death is a founding element of humanization. Scientific knowledge, then, can be acquired only at the price of a demanding and constant struggle to surmount certain obstacles—obstacles that Gaston Bachelard described as "epistemological." The first of these is personal experience, which is ascribed an undue positive value judgment. "When faced with reality, what we think we know always masks what we should know."[3] The universality of death results in the desire to provide the most general explanation possible. This promotes an unfortunately widespread tendency to base scientific culture on vast generalities. Like Bachelard, we must be wary of maxims such as "all living beings are mortal," because "we can ask ourselves whether these great laws are truly scientific thinking or, which for us is the same thing, ideas suggesting other ideas."[4]

More concretely, we can distinguish between two essential aspects of our personal experience with death. First, death seems to be a completely arbitrary fatality that is imposed on us both against our will and against our most deeply held convictions.[5] Second, death is experienced as an absurdity, probably the most intolerable of all absurdities. This results in several implicit presuppositions that until the nineteenth century and continuing today have biased our thinking on natural death. The first, and doubtless the oldest of these, says that death is programmed once and for all into the living. In ancient times, this idea was viewed as the sign of a divine power wishing to punish humanity.[6] But the idea has survived in more modern forms that lessen the arbitrary nature by bringing up physical and chemical laws. Bichat's vitalist theory, for example, attributes the fatality of death to physical forces that always end up triumphing over certain hypothetical vital forces, which by nature become depleted with age.[7] After vitalism was abandoned, the discovery of thermodynamics made it even easier to believe that physical and chemical laws were involved. Since life simply obeys the ordinary laws of the inanimate world,

[3] Bachelard, op. cit., p. 14.

[4] Bachelard, op. cit., p. 56.

[5] As stated by V. Jankélévitch in *La Mort* (Paris: Flammarion, 1977), pp. 152–153, the moment and means of death seem arbitrary and are determined only at the last minute: "the necessity of death finally ends up seeming evitable."

[6] E. Morin brings up the "archaic [idea] that death is always external, that is to say inflicted by a supernatural being or event" (*L'Homme et la Mort* [Paris: Points-Seuil, 1970], p. 331).

[7] X. Bichat, *Anatomie générale appliquée à la physiologie et à la médecine* (1801; Paris: Flammarion, 1994), p. 234.

death seems to be nothing more than the inevitable final dissolution, through the laws of entropy, of the order created by the living organism. The entropic explanation of death is commonplace, and not without interest. But it is insufficient, at least in its most common simplistic form, if only because living organisms are open systems, not in a steady state, and thus not subject to the standard laws of thermodynamics.

The fatality of death is even more evident if we consider it to be an intrinsic feature of life itself, from the beginning. Death in this case imposes itself on life from within, as part of life. Many nineteenth-century authors proposed this theory, including Lamarck and Claude Bernard, as did Metchnikoff in the early twentieth century. In a similar vein, Bichat stated that one feature of vital forces was that they grew weaker with age. In any case, whether death was in the nature of nature or in the nature of life, whether it was ordered by divine laws or by physical laws, there was simply no need to look for a biological explanation.

The second presupposition that long biased scientific thinking on death is that of a hidden utility of the function of death. This notion doubtless arose when humanity no longer blindly accepted divine will but began expecting a minimum of rationality. Although all-powerful, Linnaeus's "Sovereign Moderator," or English theologian William Paley's great "Watchmaker,"[8] would surely not have instituted death without good reason. These reasons are (more or less) accessible to the human mind, making it possible to admire the divine hand at work. In the Darwinian framework that has since been developed, these good reasons are usually expressed in terms of selective advantage. W. D. Denckla, in an article titled "A Time to Die"[9] examined "what evolutionary advantages accrue from having each adult die shortly after a few reproductive cycles." He concluded that, "while death is a tragedy for the individual, it may have evolved as a precisely timed event that increases the adaptability and hence the viability of the species." Similarly, E. Beutler stated that "mutations that are advantageous to the individual, permitting potentially eternal life, would presumably be detrimental to the species and result in its extinction."[10] To explain that species have become extinct anyway, one

[8] Who inspired R. Dawkins's *The Blind Watchmaker* (Harlow: Longman Scientific and Technical, 1986).

[9] *Life Sciences* 16 (1975): 31–44.

[10] "Planned Obsolescence in Humans and in Other Biosystems," *Perspectives in Biology and Medicine* 29 (1986):, 29, p. 175–179.

need only continue with their same line of reasoning. In the opinion of Maurice Marois, then, for whom species that disappear "suffer death for the greater good of life," and "the failure [represented by death] is surpassed by the subordination to a more vast design, subordination of individuals to the species, of species to the great and mysterious design of life."[11] This mysticism is really quite similar to the providentialism of Linnaeus.

Again, we underscore how different the situation is for the study of cell death. In cells, the phenomenon of death is not part of personal, immediate experience. On the contrary, the embryologists who first observed it in the nineteenth century needed a great deal of patience. Since the reality of cell death did not have the same force as that of the death of individuals, it was easy to dismiss as unimportant. For many years, researchers did not look into the utility of cell death despite the numerous signs pointing in this direction. The more tenuous theory of the utility of the death of individuals was much more widely accepted. How could cell death be considered a worthy research topic for the life sciences?[12] At best, it set the boundaries. This epistemological obstacle did not simply block or lead away from a research area, but it in fact prevented it from originating. According to G. Canguilhem, a similar dynamic explains why it took so long for bacteria to be considered a fundamental biological model: "How could we hope to discover the laws of the living world from the very organism—a simple parasite—that threatens it with downfall? It is doubtless because microbes were negatively valued by everyone, including biologists, in their own life experience, that they were not positively valued as objects of theoretical research [on life]."[13]

This need to find a utility for death brings to mind the similar obstacle that urges us to find a utility, real or imaginary, in our search for a reason. Bachelard noted that "the phenomena that are most hostile to man are assigned an important value whose antithetical nature would surely intrigue a psychoanalyst."[14] To accept the existence of forces that are hostile to us, it seems better to assign them a greater utility and value from the outset, rather than take the risk of studying them objectively. This com-

---

[11] In C. Chabanis, *La Mort, un terme ou un commencement?* (Paris: Fayard, 1982), p. 42.
[12] See chapter 1.
[13] Op. cit., p. 112.
[14] G. Bachelard, *La Formation de l'esprit scientifique*, 1938; 13th ed. (Paris: Vrin, 1986), p. 92.

ment applies to the conclusions of the anatomist Charles Minot, at the beginning of the twentieth century. He considered death (it is hard to find a
phenomenon more hostile to us than death!) the final, obligatory step of
the development plan that starts with the fertilization of the egg. This plan
could not exist without aging: "differentiation leads up as its inevitable
conclusion to death. Death is the price we are obliged to pay for our organization, for the differentiation which exists in us. Is it too high a price? To
that organization we are indebted for the great array of faculties with
which we are endowed [a long list follows]. And we are indebted to it also
for the possibility of the higher spiritual emotions. All of this is what we
have bought at the price of death, but it does not seem to me too much for
us to pay."[15]

Death thus seems to be useful, and even highly valued, as the toll that is
required to access the multicellular condition, and the immense advantages related to this state. This is a somewhat exalted form of the ideas of
Weismann, who also underscored the link between the loss of immortality
and the considerable enrichment represented by cell differentiation. But
Weismann, at the end of this assessment, affirmed that death did not appear on purpose, either for this reason or for any other. Far from being a
direct cost of differentiation, death is simply its fatal consequence, existing
simply because of the evolutionary mechanisms in the living world. Differentiation is what made it possible to lose immortality, not the other way
around. To use a somewhat excessive image: our first, far-removed multicellular ancestors were never required to choose between immortality and
differentiation. The appearance of differentiation marked a crucial turning point in the evolution of life. From that moment, immortality became
a cumbersome vestige, like eyes for an animal that never leaves its cave, a
vestige that had every chance of disappearing some day. By finally giving
up something that had become a useless burden, life could free up resources for more important things.

While it is true, as Bachelard said, that "the dangers of finalistic explanations have been demonstrated so often that we have no need to further
underscore how great an obstacle this is to creating a truly objective culture" (unlike in physics or chemistry), a certain form of finalism is still en-

---

[15] Quoted by B. Strehler, *Time, Cells, and Aging*, 2d ed. (New York: Academic Press, 1977).

titled to exist in biology. The danger lies in giving it too much importance, since the limits are not always easy to map out.[16] G. Williams illustrated this point with the example of the flying fish, which has to fall back into the water to prevent suffocation and death.[17] But it is not this need, however vital it may be, that explains the fish's aerodynamic shape or any of its other characteristics. The "function" of returning to the water is ensured by the laws of gravity. The parsimony principle rules out the work of natural selection in this case. Natural selection would be more likely to work in the other sense, by providing the fish with a means of remaining longer in the air. But these means are inevitably limited; they cannot change the final outcome of returning to the water. A similar reasoning applies to aging. Natural selection gives the organism the means (a notable example is nutrition) to remain alive longer than if it were abandoned only to physical forces. For various reasons—in particular, the impossibility of eliminating all risk of accident or fatal disease—these means are inevitably limited. Again, they can only delay the final outcome.

Another difficulty, more logical than psychological, hampers any attempt to explain phenomena that are even slightly complex: Namely, that we must agree on the level of explanation required. In biology, the existence of different levels of causality is particularly clear. For example, the question, "Why did Mr. Smith, 99, die during his sleep on the night of August 19th?" has several levels of responses that are not mutually exclusive but are nonetheless quite different:[18] (1) There is an internal, immediate physiological cause: Mr. Smith's heart stopped beating. This is probably what will appear on the death certificate. (2) Cardiac arrest occurred on that night, and not any other night, because the weather was particularly sultry and there was a bit of pollution in the air; Mr. Smith's cardiorespiratory system could not handle it. (3) The aging processes had progressively enfeebled Mr. Smith. If another risk factor had manifested itself at that time, for example an infection due to a flu virus, the old man would perhaps have died of that rather than of cardiac arrest. From this point of

[16] Especially because, according to Bachelard, "man does not know how to limit what is useful. That which is useful, by being highly valued, is capitalized excessively" (op. cit., quoted in P.-H. Gouyon, J. P. Henry, and J. Arnould, *Les Avatars de gène* (Paris: Belin, 1997).

[17] In *Adaptation and Natural Selection*, op. cit.

[18] This discussion is adapted from E. Mayr, *Science* 134 (1961): 1501–1506. For a very complete analysis, see A. Fagot-Largeault, *Les Causes de la mort; histoire naturelle et facteurs de risque* (Paris: Vrin, 1989).

view, this particular death can almost be considered a random event, because it was by chance that Mr. Smith encountered fatal weather conditions rather than the flu virus. (4) An in-depth analysis may show that Mr. Smith's heart was genetically predisposed to stop at around that age. Depending on each individual's genetic constitution, he or she experiences the various aging processes in a more or less pronounced manner. (5) More generally speaking, the genome of each species includes those genes that are involved in aging. The presence of these genes can be explained by the fact that natural selection has been at work since the origin of life itself. (6) If natural selection has indeed produced organisms with finite longevity, it is because sudden, extrinsic death is statistically inevitable: In such a context, it seems very likely, if not certain, that infinitely preserving the survival abilities of the soma is not the most effective strategy.

All these causes can be legitimately evoked. They can be separated into two major categories: proximal, mechanistic causes (the condition of the heart, the weather conditions, the biochemical processes that are responsible for cardiac senescence and the genes that influence these processes); and distal causes, which are based on evolution. Both types of explanations are necessary, but they must be carefully distributed. They overlap because of the complexity of the interactions between a single organism and its environment, and also because this organism is the product of a long history.

A Society Searching for Meaning

All in all, it is normal to expect biology to enhance the curative powers of medical science. The constant increase in human life expectancy has tended to transform the fight against disease into a fight against death. But the more effectively we push back death, the more absurd and intolerable it becomes. To give death meaning, it is normal to turn to biology, which has won so many other battles. This is not a new temptation. In 1915, Elie Metchnikoff, the founder of experimental gerontology, published a study on *The Death of the Silkworm Moth*, subtitled "A Chapter on Thanatology," which begins as follows: "The fact that death is often feared, like a monster we dare not look at square on, is undoubtedly one of the reasons for the ignorance of science in this area. For a person who is dying, it is not a scientist or a doctor whom we call, but a servant of the Church." These lines are even more powerful because Metchnikoff wrote them just one year before his own death. He concluded this study with the hope "that this end [without suffering and without fear of death], that our moths reach simply and naturally, will in the future be reached by humans, who will no longer need the support we mentioned at the beginning of this article." Science is supposed to prolong life to its "normal" end, to show that there is a "natural instinct for death" and also help people develop this instinct. Metchnikoff believed that our fear of death is due to the fact that too few people reach the normal end of their lives, after "accomplishing the complete physiological life cycle, with normal aging that ends with the loss of the instinct to live and the appearance of the instinct to die naturally."[19] A defunct—in the etymological sense of the word, one who has completed his or her life—should be able to approach death with serenity. We can understand why Metchnikoff subtitled his book "Essay on Optimistic Philosophy."

This quest for meaning can still be seen in many works that dismiss or reject modern evolutionary theories on aging. Macfarlane Burnet, winner of the 1960 Nobel prize for medicine, believed that "evolution and immortality are incompatible concepts. If organisms are supposed to better themselves and renew themselves every year, death is a phenomenon that is as necessary as reproduction."[20] Jean Bernard stated, "Sexual reproduction and diversity have another consequence, namely, wear, aging, and death. After producing new, different beings, the parents make room for their offspring. . . . Death is thus an advantage."[21] In the preface to one of his works published in 1987,[22] Philippe Meyer presented death as "indispensable," "a salutary event since it contains the expansion of the kingdom of the living and ensures the renewal of the species." François Jacob, winner of the 1965 Nobel prize for medicine, also seemed to share this opinion; he wrote[23] that death is a "necessary condition for the very possibility of evolution, . . . as a necessity, encoded by the genetic program into

---

[19] *Etudes sur la nature humaine*, op. cit., p. 380.

[20] *Le Programme et L'Erreur* (Paris: Albin Michel, 1982), p. 103.

[21] *Et l'âme? demande Brigitte* (Paris: Buchet-Chastel, 1987), p. 109.

[22] *Le Mythe de Jouvence* (Paris: Odile Jacob, 1987).

[23] In *La Logique du vivant*, op. cit., p. 331. However, in *Le Jeu des possibles* (Paris: Fayard, 1981), the same author presented a faithful rendition of the ideas of Medawar and Williams, which constitute the foundation for the modern theories on aging.

the egg itself," to allow "the disappearance of the generation that has fin-
ished playing its role in reproduction." And in the dictionary *Grand
Larousse Universel* (1993), the article on "death" is particularly categorical,
affirming that "the utility of death is clear: only death allows the bio-
sphere to find space on a planet that remains the same size; only death
makes biological evolution possible."

In France, this vision of death has impregnated the works of the Thanatology Society. The purpose of this society, created in 1966, is to "coordi-
nate all the branches of learning, promote and develop all research involving the problems of death, dying, and the afterlife, in all their complexity.
To this end, and with a new epistemological and pragmatic conception of
thanatology, the Society plans to found a center for the synthesis and dis-
semination of theoretical and applied knowledge on thanatological problems, specifically in the physical, chemical, somatic, ecological, sociologi-
cal, legal, psychological, moral, and religious aspects." Biology is not
mentioned per se, but the first four aspects are certainly related to science.

From 1971 to 1987, in fact, a biologist, M. Marois, was the president of
the Thanatology Society. Earlier, we mentioned his ideas on the biological
role of the death of individuals for the good of the species. He shares the
opinions of Linnaeus on another point as well, since he sees death as a
regulatory principle. He believes that along with reproduction, death ensures the balance of nature and opens the path for the evolution of species:
"A single bacterium dividing under favorable conditions, could, by geo-
metric progression, synthesize in one week a mass of living matter larger
than the mass of the earth . . . so it really must impose some sort of birth
control on itself." Marois concluded from this that "death finds its place in
the economics of life; it serves life by giving it new chances, for new attempts, for new expressions of protoplasm."[24] His ideas have had a significant influence. The anthropologist Louis-Vincent Thomas, who succeeded him as president of the Thanatology Society (from 1987 until his
death in 1994), quoted in 1975[25] statements made by Marois that were
published seven years later in a book of interviews.[26] The first traces of
Marois's influence were found in a 1971 article written by Dr. Roger Fes-
neau, then assistant secretary general of the Thanatology Society, in the

[24] In Chabanis, op. cit., p. 41.
[25] In his classic *Anthropologie de la mort* (Paris: Payot, 1975).
[26] Chabanis, op. cit.

Society Newsletter. For example, in his article "Sexuality and Death" he stated, referring to Marois, that "Death and Sexuality are two functions of life" and that "death is even indispensable to life." He concluded that "the death of the individual is therefore truly the means for preventing the death of the species." Thibault also referred to Marois when he stated that "death, which we find so revolting, so intolerable, is therefore a novelty, a selective advantage, a sort of 'progress', when looked at on an Evolutionary scale." Indeed, "the death of individuals ensures not only the perennity of the species but also its rejuvenation; it is hence not only a necessity but a *benefit*. At least in this we find a biological meaning for death (aside from any metaphysical meaning) without which it would obviously be a veritable outrage."[27] This last sentence underscores how difficult it is not to confuse biology with morality.

The same theme was taken up by J. Ruffié in his work *Le Sexe et la mort*, one of his books that best promoted the notion of the utility of death, both with the general public and in intellectual circles. He insisted in particular on the selective advantage offered by the elimination of the "old"—an argument that, as we mentioned in chapter 1, is utterly circular. His inspiration is clearly pre-Darwinian, as illustrated in his image of the predator working for the good of the prey: "Nature is constantly being cleaned by predation: hence the almost always healthy appearance of herds of wild animals. By eliminating the sick, predators decrease the chance of contagion. And by ridding the herd of the infirm and the old, predators do away with dead weight."[28] Such altruism, which is strikingly similar to the vision of Linnaeus and his disciples,[29] is diametrically opposed to the modern view of evolution and the role of natural selection. Linnaeus believed that predation in fact helped species *not to* evolve, by maintaining among them the balance desired by the "Sovereign Moderator" and by eliminating deviant individuals. Ruffié's book nonetheless caused quite a stir and gave rise to notions[30] even further removed from the scientific concepts developed during the last half century.

[27] *La Maîtrise de la mort* (Paris: Editions universitaires, 1975), pp. 15–16 (italics in the original).
[28] Op. cit., p. 226. Later he speaks about "elements that ensure the cleaning, play the role of trash collectors. All living colonies have their morticians."
[29] "The woodpecker, in pecking at rotten trees in search of insect larva, hastens their destruction so that they do not spoil the view for long. ... Dormice, by gnawing, destroy everything that is superfluous, dead, and disagreeable" op. cit., p. 115).
[30] Such as "Being born, reproducing in order not to die, dying so that others can in turn reproduce," a phrase reiterated by sociologist Annick Barrau (vice-president of the Thana-

We have no intention of denigrating multidisciplinary circles that promote reflection on death; in fact, they play an important role in reintegrating death into our Western societies, which tend to reject or hide it as a source of shame. We believe that this evolution is healthy, and the Thanatology Society doubtless contributed greatly to this change. The development of palliative care, the right to die with dignity, and the renewal of funeral practices[31] all tend to restore death to its place as an essential life event. The reader will have understood by now that our reticence and suspicion target only what we consider to be attempts to make use of biological data to create a "moral" vision of death at all costs. Moreover, the data invoked too often disregard recent information that would refute any arguments along these lines. General reflections on death should include scientific information in its rightful place, taking into account its meaning but not using it to draw conclusions that may be of some consolation but are generally merely preconceptions.

Of course, without going so far as to talk about "the great mysterious design of life" like Marois, or "resurrection through sex" like Ruffié, everyone has the right to find some comfort in the profound permanence and eternity of life.[32] But as E. Morin warned us in 1951, doing so will not pull us away from "the metaphysical meditation where we exalt the transcendent benefits of death, when we are not moaning about its no less transcendent faults [or from] myths, false proofs, and false mysteries."[33] And after all, what good is this type of consolation? V. Jankélévitch put it quite bluntly: "What compensation would the immortality of the species bring to the individual stalked by death? . . . what good does it do me that the future world will survive if I won't be there?"[34]

Biology seems to resolutely reject the idea of the utility of natural death, a utility that would allow us to find virtue in death and thus provide some consolation. Because the possibility of accidental death is never completely ruled out, the priority of living organisms cannot be to devote all of their efforts to their own survival; they must also keep resources to re-

---

tology Society in 1994–1995), in *Mort à jouer, Mort à déjouer* (Paris: Presses universitaires de France, 1994), pp. 33–34. She based her ideas on the book of J. Ruffié.

[31] Analyzed by A. Barrau, op. cit. and *Quelle mort pour demain?* (Paris: L'Harmattan, 1992).

[32] "Ever since it first appeared on earth in the lower forms, life has continued to endure without interruption," as Weismann said (op. cit.).

[33] Op. cit., p. 28.

[34] Op. cit., p. 446.

produce, to transmit their genes. Aging and natural death arise from this compromise, not automatically by a direct link but through the work of natural selection; they are simply particularly adverse side effects, from a human point of view. Thus, the constant and inevitable presence of accidental death lurking in the background has helped fashion the living, through life cycles that generally include aging and natural death.[35]

This conclusion may seem somewhat abrupt. However, the connection that each of us establishes with death is woven on a private level, through the personal, often painful awareness that we have of death, and which confers a unique value to every instant of our lives. Perhaps this is where we should turn in the quest for a "human" meaning for death—that is, depending on each person's preferences, towards philosophy or theology, rather than toward biology. But that, dear readers, is another story.

[35] It is interesting to note that, even without the support of biology, certain philosophers arrived at conclusions that echo this idea. For example V. Jankélévitch: "Life affirms itself not only despite the obstacle of death against which it protests, but also, and ipso facto, *thanks to* this obstacle . . . this obstacle that has become a condition of life, this obstacle that is at once allogenic and endogenic, never stops preventing that which it determines" (op cit., p. 97). In addition, the notion of organ-obstacle that he describes here is not unknown in biology. C. Bernard speculated that "life results from a conflict, a close, harmonic relationship between the external conditions and the predetermined conditions of the organism. It is not through a struggle against the cosmic conditions that the organism develops and maintains itself; it is on the contrary through an adaptation, an accord with them" (op. cit.). Between the organism and its environment, we find a paradoxical relationship, composed of conflict and agreement, that G. Canguilhem summed up as follows: "Living systems . . . maintain their organization both because of their openness to the outside and despite their openness" (op. cit., p. 136).

GLOSSARY

ALLELES. The different forms of a given gene present in a given species. Alleles can differ from one another by only one nucleotide. The existence of several alleles for a large number of genes translates into marked interindividual variations in most nondomesticated species. In humans, eye color and blood type are common examples of variations caused by allelic differences.

AMINO ACID. A small molecule known as the building block of proteins. There are twenty different amino acids providing a huge number of possible proteins. For a medium size protein composed of three hundred amino acids, 20^{300} different possible combinations exist.

AMYLOID PEPTIDE (Aß). In Alzheimer's disease, Aß designates the principal component of senile plaques, insoluble aggregates that are found in the brains of Alzheimer's patients and are a distinctive sign of the illness.

ANTAGONISTIC PLEIOTROPY. An evolutionary theory of aging, formulated by G. C. Williams and based on the idea that certain genes may have different effects in young and old individuals. If such a gene is beneficial to an individual early in life, enabling him or her to reproduce more effectively, it may be maintained by natural selection even if it has a harmful effect in the same individual later in life. The theory suggests that most genes involved in aging are beneficial to organisms when they are young.

APOPTOSIS (programmed cell death). Cell death that plays a physiological role, which can be considered the counterpart to cell proliferation. During embryonic development and adult life, apoptosis is involved in controlling the size of organs and eliminating superfluous or defective cells. The mechanism of apoptosis involves a series of specific genes, including caspases, which play a central role. Apoptotic cells often present a characteristic morphology known as "condensation-fragmentation." Several pathologies, such as certain types of cancer and certain neurodegenerative or autoimmune diseases, may result when apoptosis is deregulated. Apoptosis is different from necrosis, which is cell death without any physiological role or specific mechanism, and is always caused by external aggression.

ATP (adenosine triphosphate). A nucleotide that is also the energy storage molecule in cells.

CASPASES. Proteins in the protease family that play a central role in apoptosis. They are involved in an intracellular cascade triggered by pro-apoptotic signals and are responsible for destroying key proteins in cell metabolism and structure. Their active site contains the amino acid cystein, and they attack their targets after the amino acid aspartate, hence their name.

CELL CULTURE. A technique of maintaining, and often growing, animal or human cells outside the body of origin.

CELL IMMORTALITY. The unlimited ability of a cell line to multiply in culture (provided that the nutritive substrate is supplied). Immortal cells do not respect the Hayflick limit due to mutations in their genomes or chromosomal aberrations.

CELL LYSIS. The rupture of the external cell membrane (plasma membrane), resulting in cell death.

CELL SENESCENCE (proliferative senescence). A progressive loss of the ability to divide, observed in cells removed from living organisms and cultured in vitro. According to some theories, proliferative senescence constitutes the cellular basis of aging in organisms.

CHROMATIN. A supramolecular complex formed by DNA and proteins within the cell nucleus.

CHROMOSOME. A DNA molecule, often associated with proteins, which can be duplicated during cell division. The word chromosome, like chromatin, reflects this molecule's avidity for dyes, which are essential tools in histology. Dyes were instrumental in the discovery of chromosomes in the nineteenth century.

COMPLEMENTARITY (of DNA strands). This fundamental property of DNA is what enables it to be replicated—each strand acts as a template to create a new copy of the other strand during cell division. Complementarity reflects the specific pairing between nucleotides (adenine with thymine and guanine with cytosine). The same principle is used to copy the genetic information in the form of RNA during transcription. Only one of the two DNA strands acts as a template in this case.

CYTOPLASM. The main intracellular compartment, a sort of internal cell environment.

DNA (deoxyribonucleic acid). A molecule formed by a sequence of four types of small molecules (nucleotides) connected by phosphate groups. The order of the sequence determines the genetic information specific to the cell or virus containing the DNA molecule in question. DNA is generally in the form of a double molecule, the famous double helix obtained by association of two complementary strands. It can be quite long, up to tens of millions of nucleotides, but measures only a few centimeters when stretched out.

ENZYME. A specialized protein that acts as a catalyst (accelerator) for a given chemical reaction. In the absence of enzymes, these reactions would be too slow to meet the needs of a living organism.

EUKARYOTE. A cell whose DNA is in a nucleus, isolated within the cytoplasm by a membrane (as opposed to a prokaryote). By extension, an organism composed of such cells.

FREE RADICALS. An extremely unstable and reactive chemical species. They are found in cells mainly as byproducts of oxygen metabolism. Their reactivity enables them to bind and modify numerous targets, in particular DNA, proteins, and lipids involved in the composition of cell membranes. They may play a major role in the aging process. Their concentration in the organism can be decreased by using antioxidant agents.

GENE. The region of a chromosome that contains all of the information necessary to synthesize a functional molecule (protein or RNA).

GENOME. The full set of nucleic acids containing all the genes of an individual, a cell, or a virus; by extension, the information it encodes.

GERM CELLS. All reproductive cells in plants and animals (spermatozoa and ova in humans), and their precursors; as opposed to somatic cells.

HAYFLICK LIMIT. The maximum number of divisions performed by cells removed from living organisms and cultured in vitro. This limit, observed for the first time by Hayflick and Moorhead in 1959, depends on the species that the cells come from, as well as the age of the donor organism.

HOMEOTHERMIC. A "warm-blooded" animal, which maintains a constant internal temperature (as opposed to poikilothermic).

LIFE CYCLE. The combination of embryonic and postnatal development, a reproductive phase (characterized by age of first procreation and number of reproductive cycles), and often a senescent phase.

MESSENGER RNA. An RNA molecule produced by transcription of a section of DNA, which contains the information necessary to synthesize a protein. The length of the messenger RNA depends on the length of the corresponding protein, and ranges from 150 to more than 10,000 nucleotides.

MITOCHONDRIA. Cellular organelles that produce ATP, an energy storage molecule, from glucose derivatives (from food or photosynthesis in plants) and oxygen (from the atmosphere). They can be considered the respiratory organ of the eukaryotic cell. The intense oxidation-reduction processes that take place in the mitochondria create dangerous free radicals. Mitochondria have a separate genome, roughly ten to one hundred thousand times smaller than the genome in the cell nucleus, and they multiply, especially during cell division.

MUTATION ACCUMULATION THEORY. An evolutionary theory of aging in which aging is the result of the accumulation of mutations within the genome of the species during its evolution. These mutations could be retained despite natural selection because their harmful effects appear relatively late in life when the individuals have already produced descendants or have been eliminated by predators, illness, or accidents.

MUTATION. A modification of the nucleotide sequence in a genome. A mutation is transmitted to the descendants of the virus or cell where it has occurred. A mutation in a reproductive cell is called a germ mutation, which can then be transmitted to the individual's offspring. The impact of a mutation depends on its exact nature and, above all, on the functional importance of the affected region of DNA.

NEMATODE. A parasitic worm comprising the main class of nemathelminths. The species *Caenorhabditis elegans* is the most widely studied. Its small number of cells (959 somatic cells in adults) and the fact that the living animal is transparent, and can therefore be observed under a microscope, have made it an organism of choice for studying the processes involved in aging and programmed cell death.

NUCLEIC ACIDS. A general name for RNA and DNA because they are present in cell nuclei and because of the acid properties of the phosphate groups that they contain.

NUCLEOTIDE. A small molecule that is the base unit of nucleic acids. It is composed of a nitrogen base and a sugar (ribose or deoxyribose). The sugar carries at least one phosphate group. There are four nucleotides, which differ only in which base they carry: adenine, thymine (or uracil for RNA), guanine, and cytosine. The complementarity of these bases, or base pairing, is crucial for copying the genetic information during replication or transcription.

NUCLEUS. The cell compartment specific to eukaryotic cells that contains the chromosomes and is delimited by a membrane.

ONCOGENE. The gene that plays a causative role in tumor formation processes. It is often a mutant form of a gene which in its wild-type (nonmutated) form plays a physiological function generally related to cell proliferation, apoptosis, or adhesion processes. The wild-type form is called a "proto-oncogene."

PARTHENOGENESIS. An asexual form of reproduction from a nonfertilized egg. It occurs naturally in many social insects and can be induced artificially in other species.

PLEIOTROPY (OR PLEIOTROPISM). A word created by geneticists to describe the involvement of a single gene in several functions, or more generally in several characteristics of an organism.

POIKILOTHERMIC. A "cold-blooded" animal, or etymologically "variable temperature" depending on the temperature of the environment (as opposed to homeothermic).

PROKARYOTE. A cell whose genome is not isolated from the cytoplasm by a membrane that delimits a nucleus (as opposed to a eukaryotic cell).

PROTEASES. A class of enzymes involved in the breakdown of proteins.

PROTEIN. A molecule formed by a chain of many amino acids, in an order determined by the corresponding gene. A single cell can produce up to several thousand different proteins, including catalysts for chemical reactions (enzymes), proteins that are receptors for external signals such as hormones and neurotransmitters, and proteins that give the cell its shape and motility.

RIBOSOMAL RNA. RNA involved in the formation of ribosomes.

RIBOSOMES. The supramolecular assembly of proteins and RNA, on which the synthesis of proteins from messenger RNA is performed. This process is also known as translation.

RNA (ribonucleic acid). Very similar in composition and structure to DNA, RNA plays several essential roles in the expression of genetic information (see messenger RNA and ribosome). RNA is made by copying DNA. More than 80 percent of a cell's RNA is found in the ribosomes. Unlike DNA, RNA is usually single-strand.

SEQUENCE. The order of the nucleotides in a gene or of amino acids in a protein. It is a type of higher order chemical formula.

SOMATIC CELLS. All nonreproductive cells, as opposed to germ cells.

TELOMERE (telomerase). The end of chromosomes. According to certain theories, the progressive shortening of the telomeric ends during cell division is at the origin of cell senescence in somatic cells. In germ cells, the shortening is prevented by the activity of a specialized enzyme—telomerase.

TRANSCRIPTION. The synthesis of a molecule of RNA from a gene by complementarity with one of the two strands of the corresponding DNA region.

TRANSGENESIS. The technology used to insert additional genes into the genome of an animal or plant. A related technology, "homologous recombination," is used to eliminate or modify certain genes. This technique is used to create new characters in the modified animal to provide a better understanding of the function of the manipulated genes.

TRANSGENIC ANIMAL. See transgenesis.

TRANSLATION. The process of assembling a protein from the genetic information contained in messenger RNA. Ribosomes play a major role in translation.

TUMOR SUPPRESSOR GENE (anti-oncogene). The gene that encodes a product involved in the control and repression of cellular malfunctions that may lead to the formation of a tumor. A mutation in a tumor suppressor gene can be one cause of cancer (as in the case of the *p53* gene).

BIBLIOGRAPHY

This is not an exhaustive list of the references for the findings reported in this book, but a collection of books and journal articles that will give the reader an overview of the topics discussed.

General Works

Allard, M. *A la recherche du secret des centenaires.* Paris: Le Cherche Midi, 1991.

Ameisen, J. C. *La Sculpture du vivant.* Paris: Seuil, 1999.

Bachelard, G. *La Formation de l'esprit scientifique.* 13th ed. 1938. Reprint, Paris: Vrin, 1986.

Barrau, A. *Quelle mort pour demain?* Paris, L'Harmattan, 1992.

——. *Mort à jouer, mort à déjouer.* Paris: Presses universitaires de France, 1994.

Bell, G. *Sex and Death in the Protozoa: The History of an Obsession.* Cambridge: Cambridge University Press, 1988.

Bernard, C. *Leçons ser les phénomènes de la vie communs aux végétaux et aux animaux.* 1878. Reprint, Paris: Buchet-Chastel, 1987.

Bichat, X. *Anatomie générale appliquée à la physiologie et à la médecine.* 1801. Reprint, Paris: Flammarion, 1994.

——. *Physiological Researches on life and death.* 1800. Translated by F. Gold. New York: Arno, 1997.

Canguilhem, G. *Ideology and Rationality in the History of the Life Sciences.* Translated by A. Goldhammer. Cambridge: MIT Press, 1988.

Chabanis, C. *La Mort, un terme ou un commencement?* Paris: Fayard, 1982.

Dawkins, R. *The selfish gene.* Oxford: Oxford University Press, 1976.

Fagot-Largeault, A. *Les Causes de la mort, histoire naturelle et facteurs de risque.* Paris: Vrin, 1989.

Finch, C. E. *Longevity, Senescence, and the Genome.* Chicago: University of Chicago Press, 1990.

Gouyon, P.-H., J.-P. Henry and J. Arnould. *Gene Avatars: The Neo-Darwinian Theory of Evolution.* Translated by T. Ojasoo. New York: Kluwer Academic, 2002.

Hayflick, L. *How and Why We Age.* New York: Ballantine, 1994.

Jacob, F. *Le Jeu des possibles.* Paris: Fayard, 1981.

——. *The Logic of Life: A History of Heredity.* Translated by B. E. Spillmann. 1974. Reprint, with a new preface by the author, Princeton: Princeton University Press, 1993.

Jankélévitch, V. *La Mort.* Paris: Flammarion, 1977.

Lamarck, J.-B. *Zoological Philosophy: An Exposition with Regard to the Natural History of Animals.* 1809. Translated by H. Elliot. Chicago: University of Chicago Press, 1984.

——. *Recherches sur l'organisation des corps vivants.* 1902. Reprint, Paris: Flammarion, 1986.

Langaney, A. *La Philosophie biologique.* Paris: Belin, 1999.

Linnaeus, Carl. *L'Equilibre de la nature.* 1744–1760. Translated by B. Jasmin, with introduction and notes by C. Limoges. Paris: Vrin, 1972.

Mayr, E. *The Growth of Biological Thought: Diversity, Evolution, and Inheritance.* Cambridge: Harvard University Press, 1982.

Metchnikoff, E. *The Nature of Man: Studies in Optimistic Philosophy.* Translation edited by P. C. Mitchell. 1903. Reprint, New York: Arno, 1977.

Meyer, P. *Le Mythe de Jouvence.* Paris: Odile Jacob, 1987.

Morange, M. *The Misunderstood Gene.* Translated by M. Cobb. Cambridge: Harvard University Press, 2001.

Morin, E. *L'Homme et la Mort.* 1951. 2d ed. Paris: Points-Seuil, 1970.

Pichot, A. *Histoire de la notion de vie.* Paris: Gallimard, 1993.

Prochiantz, A. *Claude Bernard, la révolution physiologique.* Paris: Presses universitaires de France, 1990.

Rostand, J. *Esquisse d'une histoire de la biologie.* Paris: Gallimard, 1945.

Ruffié, J. *Le Sexe et la Mort.* Paris: Odile Jacob, 1986.

Strehler, B. *Time, Cells, and Aging.* 1962. 2d ed. New York: Academic Press, 1977.

Thomas, L.-V. *Anthropologie de la mort.* Paris: Payot, 1975.

Weismann, A. "The Duration of Life." 1881. In his *Essays upon Heredity and Kindred Biological Problems.* Oxford: Oxford University Press, 1889.

——. "Life and Death." 1883. In his *Essays upon heredity and kindred biological problems.* Oxford: Oxford University Press, 1889.

Williams, G. C. *Adaptation and Natural Selection.* Princeton: Princeton University Press, 1966.

"Vivre 120 ans." Special issue of *La Recherche,* July–August 1999.

More Specialized Articles or Works

Bulmer, M. *Theoretical Evolutionary Ecology.* Sunderland, Mass.: Sinauer Associates, 1994.

Clarke, P. G. H., and S. Clarke. "Nineteenth Century Research on Naturally Occurring Cell Death and Related Phenomena." *Anat. Embryol.* 193 (1996): 81–99.

Evan, G., and T. Littlewood. "A Matter of Life and Cell Death." *Science* 281 (1998): 1317–1322.

Hamburger, V. "The History of the Discovery of Neuronal Death in Embryos." *Journal of Neurobiology* 23 (1992): 1116–1123.

Jaworowski, A., and S. M. Crowe. "Does HIV Cause Depletion of CD4 + T Cells in Vivo by the Induction of Apoptosis?" *Immunology and Cell Biology* 77 (1999): 90–98.

Johnson, F., D. Sinclair, and L. Guarente. "Molecular Biology of Aging." *Cell* 96 (1999): 291–302.

Kirkwood, T., and T. Cremer. "Cytogerontology since 1881: A Reappraisal of August Weismann and a Review of Modern Progress." *Human Genetics* 60 (1982): 101–121.

Kirkwood, T., and R. Holliday. "The Evolution of Ageing and Longevity." *Philos. Trans. Roy. Soc. Lond. B* 332 (1979): 15–24.

Levine, A. "p53, the Cellular Gatekeeper from Growth and Division." *Cell* 88 (1997): 323–331.

Martin, G. M. "Genetics and the Pathobiology of Ageing." *Philos. Trans. Roy. Soc. Lond. B Bio. Sci.* 352 (1997): 1773–1780.

Mayr, E. "Cause and Effect in Biology." *Science* 134 (1961): 1501–1506.

Medawar, P. B. *An Unsolved Problem of Biology.* London: K. K. Lewis, 1952.

Metzstein, M. M., G. M. Standfield, and H. R. Horvitz. "Genetics of Programmed Cell Death in *C. elegans:* Past, Present and Future." *Trends in Genetics* 14 (1998): 410–416.

Partridge, L., and N. H. Barton. "Optimality, Mutation and the Evolution of Ageing." *Nature* 362 (1993): 305–311.

Raff, M. "Cell Suicide for Beginners." *Nature* 396 (1998): 119–122.

Rose, M. *Evolutionary Biology of Aging.* Oxford: Oxford University Press, 1991.

Sedivy, J. M. "Can Ends Justify the Means? Telomeres and the Mechanisms of Replicative Senescence and Immortalization in Mammalian Cells." *Proc. Nat. Acad. Sci. USA* 95 (1998): 9078–9081.

Smith, J. R., and O. M. Pereira-Smith. "Replicative Senescence: Implications for in Vivo Aging and Tumor Suppression." *Science* 273 (1996): 63–67.

Sohal, R. S., and R. Weindruch. "Oxidative Stress, Caloric Restriction, and Aging." *Science* 273 (1996): 59–63.

Toussaint, O., and J. Remacle. "Revue critique des théories du vieillissement cellulaire du concept de base de Hayflick au concept de seuil critique d'accumulation d'erreurs." *Pathologie Biologie* 42 (1994): 313–321.

Wallace, D. C. "Mitochondrial Diseases in Man and Mouse." *Science* 283 (1999): 1482–1488.

Williams, G. C. "Pleiotropy, Natural Selection, and the Evolution of Senescence." *Evolution* 11 (1957): 398–411.

See also a collection of review articles on "Ageing," *Nature* 408 (2000): 230–269.

INDEX